PHILIP'S BRITAIN & IRELAND

2019

STARGAZING

MONTH-BY-MONTH GUIDE TO THE NIGHT SKY

HEATHER COUPER & NIGEL HENBEST

www.philipsastronomy.com
www.philips-maps.co.uk

Published in Great Britain in 2018 by Philip's,
a division of Octopus Publishing Group Limited
(www.octopusbooks.co.uk)
Carmelite House, 50 Victoria Embankment,
London EC4Y 0DZ
An Hachette UK Company (www.hachette.co.uk)

TEXT
Heather Couper and Nigel Henbest © 2018 pages 4–85
CPRE © 2017 pages 90–91
Philip's © pages 2018 pages 1–3
Robin Scagell © 2018 pages 86–89

MAPS
pages 92–95 © OpenStreetMap contributors, Earth
Observation Group, NOAA National Geophysical Data
Center. Developed by CPRE and LUC.

ARTWORKS © Philip's

ISBN 978-1-84907-480-3

A CIP catalogue record for this book is available from the
British Library.

Printed in China

CONTENTS

Welcome to the magical world of *Stargazing*! Within these pages, you'll find your complete guide to everything that's happening in the night sky throughout 2019 – whether you're a beginner or a seasoned astronomer.

With the 12 monthly star charts, you can find your way around the sky on any night in the year. Impress your friends by identifying celestial sights ranging from the brightest planets to some pretty obscure constellations.

Every page of *Stargazing 2019* is bang up to date, bringing you up to speed with everything that's new this year, from shooting stars to eclipses. And we'll start with a run-down of the most exciting sky-sights on view this year (opposite).

THE MONTHLY CHARTS

A reliable map is as essential for exploring the heavens as it is for visiting another country. For each month, we provide a circular **star chart** showing the whole evening sky. To keep the maps uncluttered, we've plotted about 200 of the brighter stars (down to third magnitude), which means you can pick out the main star patterns – the constellations. (If we'd shown every star visible on a dark night, there'd be around 3000 stars on the charts!) We also show the ecliptic: the apparent path of the Sun, closely followed by the Moon and planets as well.

You can use these charts throughout the UK and Ireland, along with most of Europe, North America and northern Asia – between 40 and 60 degrees north – though our detailed timings apply specifically to the UK and Ireland.

USING THE STAR CHARTS

It's pretty easy to use the charts. Start by working out your compass points.

South is where the Sun is highest in the sky during the day; east is roughly where the Sun rises, and west where it sets. At night, you can find north by locating the Pole Star – Polaris – by using the stars of the Plough (see March).

The left-hand chart then shows your view to the north. Most of the stars here are visible all year: these circumpolar constellations wheel around Polaris as the seasons progress.

Your view to the south appears in the right-hand chart; it changes much more as the Earth orbits the Sun. Leo's prominent 'sickle' is high in the spring skies. Summer is dominated by the bright trio of Vega, Deneb and Altair. Autumn's familiar marker is the Square of Pegasus; while the stars of Orion rule the winter sky.

During the night, our perspective on the sky also alters as the Earth spins around, making the stars and planets appear to rise in the east and set in the west. The charts depict the sky in the late evening (the exact times are noted in the captions). As a rule of thumb, if you are observing two hours later, then the following month's map will be a better guide to the stars on view – though beware: the Moon and planets won't be in the right place!

THE PLANETS, MOON AND SPECIAL EVENTS

Our charts also highlight the **planets** above the horizon in the late evening. We've indicated the track of any **comets** known at the time of writing; though we're

HIGHLIGHTS OF THE YEAR

- At the **beginning of January**, you may catch Comet Wirtanen with the naked eye, high in the northern sky.
- **Early January** sees the best appearance of Venus as the Morning Star.
- **Night of 3/4 January:** the Quadrantid meteor shower, unspoilt by moonlight, will be the most spectacular shooting star display of 2019.
- **Night of 20/21 January:** in the early hours, we're treated to the most stunning eclipse we'll see from the UK this year, as a supermoon is engulfed by the Earth's shadow.
- **22 January:** Venus close to Jupiter in the morning sky.
- **Night of 13/14 February:** the Moon moves through the Hyades star cluster, near Aldebaran.
- **18 February:** Venus passes close to Saturn in the dawn twilight.
- **19 February:** the biggest and brightest supermoon of 2019 will be the most spectacular for the next eight years.
- **30 March:** Mars passes close to the Pleiades (Seven Sisters).
- **13 April:** the Moon moves right in front of the Beehive Cluster (Praesepe).
- **Night of 5/6 May:** excellent views of the Eta Aquarid meteor shower – tiny pieces shed by Halley's Comet.
- **10 June:** Jupiter is opposite to the Sun in the sky, and at its brightest this year.
- **18 June:** Mercury very close to Mars.
- **2 July:** a total eclipse of the Sun is visible from parts of Chile and Argentina.
- **9 July:** Saturn is opposite to the Sun in the sky and at its closest to Earth this year.
- **16 July:** a 65 per cent partial eclipse of the Moon, visible from the British Isles.
- **Night of 12/13 August:** maximum of the prolific Perseid meteor shower, though the fainter shooting stars are washed out by moonlight this year.
- **Night of 23/24 August:** the Moon hides several stars in the Hyades.
- **6 September:** Neptune lies almost in front of the star phi Aquarii.
- **17 October:** the Moon swims through the Hyades just above Aldebaran.
- **Morning of 22 October:** the Moon moves in front of Praesepe (the Beehive cluster).
- **11 November:** Mercury transits across the face of the Sun. It's the last time we'll see this rare phenomenon until 2032.
- **Night of 13/14 November:** the Moon occults many stars in the Hyades.
- **24 November:** the two brightest planets, Venus and Jupiter, are close together in the south-west after sunset.
- **10, 11 December:** Brilliant Venus passes Saturn.
- **Nights of 13/14 and 14/15 December:** peak of the spectacular Geminid meteor shower, though moonlight will take the edge off the show.
- **26 December:** an annular eclipse of the Sun is visible from a narrow path through the Arabian peninsula, southern India and Sri Lanka to Indonesia and Singapore. None of the eclipse is visible from the British Isles.

afraid we can't guide you to a comet that's found after the book has been printed!

We've plotted the position of the Full Moon each month, and also the **Moon's position** at three-day intervals before and afterwards. If there's a **meteor shower** in the month, we mark its radiant – the position from which the meteors stream.

The **Calendar** provides a daily guide to the Moon's phases and other celestial happenings. We've detailed the most interesting in the **Special Events** section, including close pairings of the planets, times of the equinoxes and solstices and – most exciting – **eclipses** of the Moon and Sun.

Check out the **Planet Watch** page for more about the other worlds of the Solar System, including their antics at times they're not on the main monthly charts. We've illustrated unusual planetary, lunar and cometary goings-on in the **Planet Event Charts**. And there's a full explanation of all these events in the **Solar System 2019** on pages 80–82.

MONTHLY OBJECTS, TOPICS AND PICTURES

Each month, we examine one particularly interesting **object**: a planet, perhaps, or a star or a galaxy. We also feature a spectacular **picture** – taken by a backyard amateur based in Britain – and describe how the image was captured. And we explore a fascinating and often newsworthy **topic**, ranging from the naming of exoplanets to gravitational waves.

GETTING IN DEEP

There's a practical **observing tip** each month, helping you to explore the sky with the naked eye, binoculars or a telescope.

New for this year is our guide to the **Top 20 Sky Sights**, such as nebulae, star clusters and galaxies. You'll find it on pages 83–85.

For a round-up of the latest in **observing technology**, turn to pages 86–89, where equipment expert Robin Scagell offers advice on how to choose your first telescope. And check out the **dark-sky maps** (pages 90–95), showing where to find the blackest skies in Great Britain, and enjoy the most breathtaking views of the heavens.

So: fingers crossed for good weather, glorious eclipses, a multitude of meteors and – the occasional surprise.

Happy stargazing!

JARGON BUSTER

Have you ever wondered how astronomers describe the brightness of the stars, or how far apart they appear in the sky? Not to mention how to measure the distances to the stars. If so, you can quickly find yourself mired in some arcane astro-speak – magnitudes, arcminutes, light years and the like.

Here's our quick and easy guide to busting that jargon:

Magnitudes

It only takes a glance at the sky to see that some stars are pretty brilliant, while many more are dim. But how do we describe to other people how bright a star appears?

Around 2000 years ago, ancient Greek astronomers ranked the stars into six classes, or **magnitudes**, depending on their brightness. The most brilliant stars were first magnitude, and the faintest stars you can see came in at sixth magnitude. So the

stars of the Plough, for instance, are second magnitude while the individual Seven Sisters in the Pleiades are fourth magnitude.

Today, scientists can measure the light from the stars with amazing accuracy, so we can state their brilliance with more precision. (Mathematically speaking, a difference of five magnitudes represents a difference in brightness of 100 times.) So the Pole Star is magnitude +2.02, while Vega is magnitude +0.03. Because we've inherited the ancient ranking system, the brightest stars have the *smallest* magnitude. In fact, the brightest stars come in with a negative magnitude, including Sirius (magnitude –1.47).

And we can use the magnitude system to describe the brightness of other objects in the sky, such as stunning Venus which can be almost as brilliant as magnitude –5. The Full Moon and the Sun have whopping negative magnitudes!

Venus (centre) at magnitude –4.7 is just over a hundred times brighter than Aldebaran (the red star below) at magnitude +0.9.

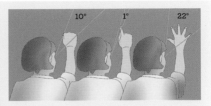

At the other end of the scale, stars, nebulae and galaxies with a magnitude fainter than +6.5 are too dim to be seen by the naked eye. Using ever larger telescopes – or by observing from above Earth's atmosphere – you can perceive fainter and fainter objects. The most distant galaxies visible to the Hubble Space Telescope are ten billion times fainter than the naked eye limit.

Here's a guide to the magnitude of some interesting objects (quoted to the nearest tenth of a magnitude):

Sun	–26.7
Full Moon	–12.5
Venus (at its brightest)	–4.7
Sirius	–1.5
Betelgeuse	+0.4
Polaris (Pole Star)	+2.0
Faintest star visible to the naked eye	+6.5
Faintest star visible to the Hubble Space Telescope	+31

Degrees of separation

Astronomers measure the distance between objects in the sky in **degrees** (symbol °): all around the horizon is 360°, while it's 90° from the horizon to the point directly overhead (the zenith).

As we show in the diagram, you can use your hand – held at arm's length – to give a rough idea of angular distances in the sky.

For objects that are very close together – like many double stars – we divide the degree into 60 arcminutes (symbol '). And for celestial objects that are very tiny – such as the discs of the planets – we split each arcminute into 60 arcseconds (symbol "). To give you an idea of how small these units are, it takes 3600 arcseconds to make up one degree.

Here are some typical separations and sizes in the sky:

Length of the Plough	25°
Width of Orion's Belt	3°
Diameter of the Moon	31'
Separation of Mizar and Alcor	12'
Diameter of Jupiter	45"
Separation of Albireo A and B	35"

How far's that star?

The Universe is a big place, and everything we see in the heavens lies a long way off. We can give distances to the planets in millions of kilometres. But the stars are so distant that even the nearest, Proxima Centauri, lies some 40 million million kilometres away. To turn those distances into something more manageable, astronomers use a larger unit: one **light year**.

A light year is about 9.46 million million kilometres – the distance that light travels in a year. That makes Proxima Centauri a much more manageable 4.2 light years away from us. Here are the distances to some other familiar astronomical objects, in light years:

Sirius	8.6
Betelgeuse	640
Centre of the Milky Way	27,000
Andromeda Galaxy	2.5 million
Most distant galaxies seen by Hubble Space Telescope	13 billion

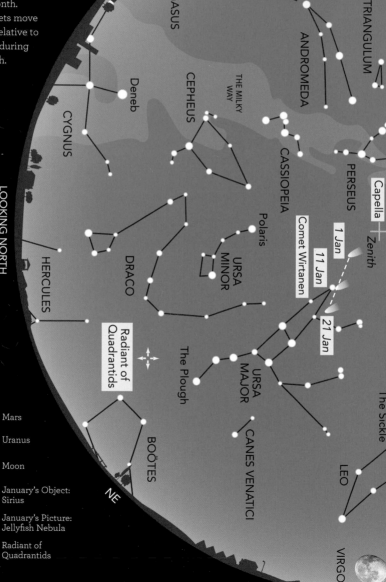

WEST

- The sky at 10 pm in mid-January, with Moon positions at three-day intervals either side of Full Moon.
- The star positions are also correct for 11 pm at the beginning of January, and 9 pm at the end of the month.
- The planets move slightly relative to the stars during the month.

NW

LOOKING NORTH

NE

PISCES

Mars

Square of Pegasus

TRIANGULUM

PEGASUS

ANDROMEDA

THE MILKY WAY

Deneb

CEPHEUS

CASSIOPEIA

PERSEUS

Capella

Zenith

AURIGA

CYGNUS

Polaris

Comet Wirtanen

1 Jan

11 Jan

HERCULES

DRACO

URSA MINOR

21 Jan

Radiant of Quadrantids

The Plough

URSA MAJOR

The Sickle

BOÖTES

CANES VENATICI

LEO

VIRGO

24 Jan

Mars

Uranus

Moon

January's Object: Sirius

January's Picture: Jellyfish Nebula

Radiant of Quadrantids

EAST

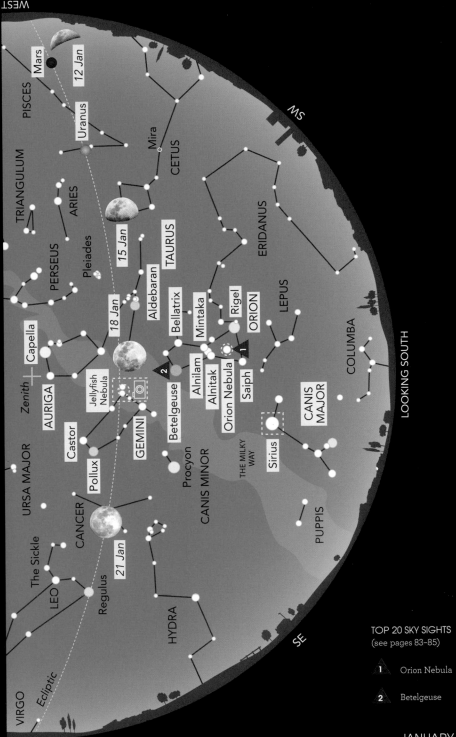

JANUARY

Mars

12 Jan

PISCES

Uranus

MS

TRIANGULUM

Mira

CETUS

ARIES

PERSEUS

15 Jan

TAURUS

ERIDANUS

Pleiades

Aldebaran

Bellatrix

Mintaka

Rigel

ORION

18 Jan

Capella

LEPUS

Zenith

AURIGA

Jellyfish Nebula

Alnilam

Alnitak

Orion Nebula

Saiph

2

1

COLUMBA

Castor

Betelgeuse

GEMINI

CANIS MAJOR

URSA MAJOR

Pollux

Procyon

Sirius

THE MILKY WAY

LOOKING SOUTH

CANCER

CANIS MINOR

The Sickle

21 Jan

PUPPIS

LEO

Regulus

SE

HYDRA

VIRGO

Ecliptic

TOP 20 SKY SIGHTS
(see pages 83–85)

1   Orion Nebula

2   Betelgeuse

The year begins with a cornucopia of celestial sights, from a comet to colourful shooting stars, from a glorious Morning Star to an eclipse of the supermoon. Plus a bevy of brilliant stars: **Betelgeuse** and **Rigel** in **Orion**; **Aldebaran**, the bright red eye of **Taurus** (the Bull); **Capella**, crowning **Auriga** (the Charioteer); **Castor** and **Pollux**, the celestial twins in **Gemini**; and glorious **Sirius**, in **Canis Major** (the Great Dog).

JANUARY'S CONSTELLATION

Spectacular **Orion** is a rare star grouping that looks like its namesake – a giant hunter with a sword below his belt, wielding a club. The seven main stars lie in the 'top 70' brightest stars in the sky, but they're not closely associated – they simply line up, one behind the other.

Closest – at 250 light years – is the fainter of the two stars forming the hunter's shoulders, **Bellatrix**. Next is the other shoulder-star, blood-red **Betelgeuse**. This giant star, 640 light years away, is a thousand times larger than our Sun, and its fate will be to explode as a supernova (see December's Object).

Slightly brighter, blue-white **Rigel** (Orion's foot) – 860 light years from us – is a young star twice as hot as our Sun, and 125,000 times more luminous. **Saiph**, the hunter's other foot, is 650 light years distant. The two outer belt stars, **Alnitak** and **Mintaka** lie 700 and 690 light years away.

We travel 1300 light years from home to reach the middle star of the belt, **Alnilam**, and the stars of the 'sword' hanging below – the lair of the great **Orion Nebula**. Some 24 light years across, it's the nearest massive 'star factory'.

JANUARY'S OBJECT

Sirius is lording it over the night skies this month, at the head of **Canis Major** (the Great Dog). Though it's our brightest star, at magnitude –1.47, Sirius is not particularly luminous: it just happens to lie nearby, at 8.6 light years.

Popularly known as 'the Dog Star', it was named Seirios ('scorcher') by the ancient Greeks. They believed it added to the Sun's heat in summer to bring the hot and humid 'dog days', when everything – including canines – slowed down. To the Egyptians, the summer appearance of Sirius heralded the Nile floods – a welcome harbinger of a bumper harvest.

Boasting a temperature of about 10,000 °C, Sirius is twice as heavy as the Sun. And it's relatively young: just 230 million years old, as compared to the Sun's venerable 4,600 million years.

OBSERVING TIP

Hold a 'meteor party' to check out the spectacular Quadrantid meteor shower on 3/4 January. The ideal viewing equipment comprises just your unaided eyes, plus a warm sleeping bag and a lounger. Everyone should look in different directions, to cover the whole sky: shout out 'Meteor!' when you see a shooting star. One person can record the observations, using a watch, notepad and red torch. In the interests of science, try to observe the sky for at least an hour, before repairing indoors for some warming meteoric celebrations....

With a 150 mm (or larger) telescope, you can spot a companion star ten thousand times fainter, at magnitude +8.4. Nicknamed 'the Pup', it was once heavier than Sirius, but it has now puffed off its outer layers. The same weight as the Sun, and yet only the size of the Earth, the Pup is a dense white dwarf, with a searing surface temperature of 25,000°C and considerable gravitational powers. But it's on the road to oblivion. With no nuclear reactions, the Pup will cool to become a dead, black globe. Just wait two billion years...

JANUARY'S TOPIC: KUIPER BELT ENCOUNTER

After its spectacular fly-by of the dwarf planet Pluto in 2015, NASA's spacecraft New Horizons is encountering a new world, recently named Ultima Thule by the public, on NASA's invitation. 'Mr Pluto' – Alan Stern – enthuses: 'Spend the New Year's Eve with NASA – in the **Kuiper Belt**'.

Predicted by Dutch-American astronomer Gerard Kuiper, the Belt is a swarm of bodies like the asteroid belt, except that there are more of them, and they're icy, not rocky. These 'KBOs' are leftover debris from our Solar System's birth, providing clues to our origins.

Ultima Thule measures just 30 km across, and it appears to be double. It's also the most distant object that a spacecraft has explored. What can we expect?

Surprises! Even in the extreme cold of the outer Solar System, KBOs are awesomely active. Take Pluto, with its ice volcanoes and nitrogen glaciers.

And the excitement isn't over yet. New Horizons has enough fuel to operate well into the 2030s – so watch this space as it investigates these far-flung mini-worlds!

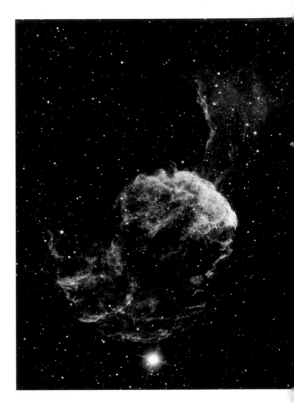

Pete Williamson used a T20 106 mm f/5 Takahashi refractor to capture the Jellyfish. He took nine 300-second exposures through three filters: sulphur, hydrogen and oxygen.

JANUARY'S PICTURE

Some 5000 light years away in Gemini, the **Jellyfish Nebula** (IC 443) is the corpse of a supernova that exploded 30,000 years ago. Its core survives as a pulsar, a rapidly spinning neutron star.

The Jellyfish is 70 light years across, and appears larger than the Moon. A telescope reveals its delicate filaments of gas, where the expanding nebula is squeezing gases to create new stars. Each is a cosmic phoenix, rising from the ashes of the exploded star.

SUNDAY	MONDAY	TUESDAY	WEDNESDAY	THURSDAY	FRIDAY	SATURDAY
		1 New Horizons at Ultima Thule; Moon near Venus (am)	**2** Crescent Moon with Venus and Jupiter (am)	**3** Moon near Jupiter (am); Earth at perihelion	**4** Quadrantids (am)	**5**
6 1.28 am New Moon; solar eclipse; Venus W elongation	**7**	**8**	**9**	**10**	**11**	**12** Moon near Mars
13	**14** 6.46 am First Quarter Moon	**15**	**16**	**17** Moon very close to Aldebaran	**18**	**19**
20	**21** 5.16 am Full Moon; supermoon; lunar eclipse	**22** Venus near Jupiter (am); Moon near Regulus	**23**	**24**	**25**	**26**
27 Moon near Spica (am); 9.11 pm Last Quarter Moon	**28**	**29**	**30** Moon near Venus and Jupiter (am)	**31** Moon between Venus and Jupiter (am)		

SPECIAL EVENTS

• At the start of the month, you may catch **Comet Wirtanen** – high in the northern sky, in Ursa Major (see Star Chart) – with the naked eye. It's fading rapidly, so after mid-January you'll need binoculars to spot it.

• **1 January:** The New Horizons spacecraft passes Ultima Thule (see Topic).

• **1–3 January:** the crescent Moon passes Venus and Jupiter in the dawn sky.

• **3 January, 5.20 am:** the Earth is closest to the Sun (147 million km away).

• **Night of 3/4 January:** watch after midnight for the most spectacular shooting star display of 2019! There's no moonlight to interfere with the **Quadrantid meteor shower,** dust particles from the old comet 2003 EH₁ burning up in the Earth's atmosphere.

• **6 January:** a partial eclipse of the Sun is visible from the northwest Pacific, including Japan and part of China.

• **17 January:** the star immediately below the Moon is Aldebaran (Chart 1a).

• **Night of 20/21 January:** in the early hours, we're treated to the most stunning eclipse we'll see from the UK this year, as an unusually large and bright supermoon is engulfed by the Earth's shadow. Visible in the Americas, western Africa and Europe, from the UK, the eclipse begins at 3.34 am; totality lasts from 4.41 to 5.43 am.

• **30, 31 January:** around 6 am, the crescent Moon joins Venus, Jupiter and Antares (Chart 1b).

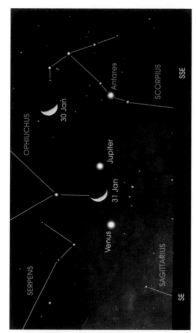

1a 17 January, 8 pm. The Moon lies close to Aldebaran.

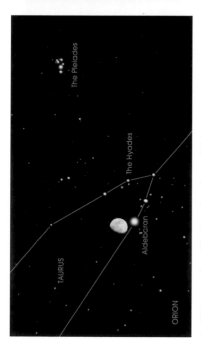

1b 30-31 January, 6 am. The crescent Moon passes Jupiter and Venus.

- **Mars** is the only prominent planet in the evening sky. Shining at magnitude +0.7 in the dim constellation Pisces, the Red Planet sets around 11.30 pm.
- With binoculars or a small telescope, you'll find **Neptune** (magnitude +7.9) in Aquarius, setting at about 9 pm; and **Uranus** to the upper left of Mars in Pisces, with a magnitude of +5.8 and disappearing below the horizon at around 1 am.
- Most of the planetary action is taking place in the hours before dawn. Brilliant **Venus** rises about 4.30 am, blazing at magnitude −4.4 as it travels from Libra, through Scorpius to Ophiuchus.
- **Jupiter** lies in Ophiuchus near Antares, the principal star of Scorpius. At the beginning of January, the giant world is rising at 6 am, to the lower left of Venus. The second-brightest planet (though still outshining any star) Jupiter is ten times fainter than Venus, at magnitude −1.8. As the month progresses, the two planets move closer together, with Venus passing over Jupiter on the morning of 22 January. Before dawn on 30 and 31 January, there's a lovely sight as the crescent Moon joins Jupiter, Venus and Antares (Chart 1b).
- Right at the end of the month, you may catch **Saturn** rising at 7 am in the dawn twilight, to the lower left of Venus and Jupiter. It shines at magnitude +0.6 in Sagittarius.
- **Mercury** is too close to the Sun to be visible in January.

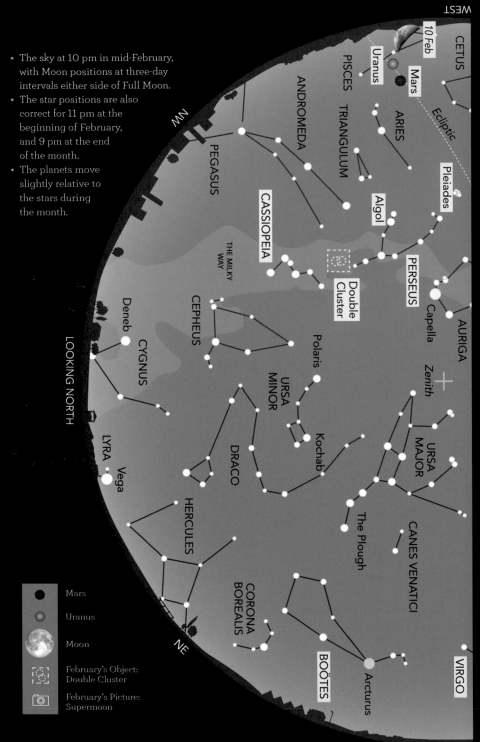

- The sky at 10 pm in mid-February, with Moon positions at three-day intervals either side of Full Moon.
- The star positions are also correct for 11 pm at the beginning of February, and 9 pm at the end of the month.
- The planets move slightly relative to the stars during the month.

WEST

10 Feb

CETUS

Uranus

Mars

PISCES

Ecliptic

TRIANGULUM

ANDROMEDA

ARIES

Pleiades

PEGASUS

Algol

PERSEUS

CASSIOPEIA

Double Cluster

THE MILKY WAY

Capella

AURIGA

Zenith

Deneb

CEPHEUS

NW

LOOKING NORTH

CYGNUS

Polaris

URSA MINOR

URSA MAJOR

LYRA

Vega

DRACO

Kochab

The Plough

CANES VENATICI

HERCULES

CORONA BOREALIS

BOÖTES

Arcturus

VIRGO

NE

Legend:

- ● Mars
- ◉ Uranus
- ○ Moon
- February's Object: Double Cluster
- February's Picture: Supermoon

EAST

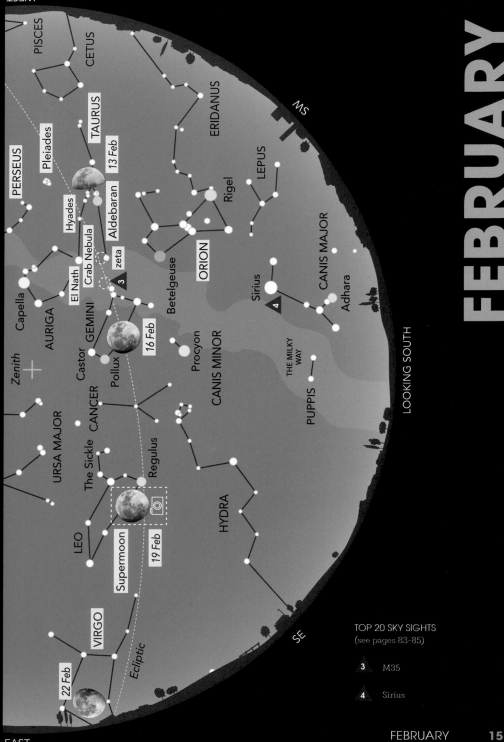

WEST

PISCES

CETUS

TAURUS

Pleiades

PERSEUS

13 Feb

Hyades

Aldebaran

El Nath

Crab Nebula

zeta

Capella

AURIGA

GEMINI

Castor

Pollux

16 Feb

Zenith

URSA MAJOR

CANCER

The Sickle

Regulus

LEO

Supermoon

19 Feb

VIRGO

22 Feb

Ecliptic

ERIDANUS

SW

LEPUS

Rigel

ORION

Betelgeuse

Procyon

CANIS MINOR

HYDRA

Sirius

CANIS MAJOR

Adhara

THE MILKY WAY

PUPPIS

SE

LOOKING SOUTH

3 M35

4 Sirius

TOP 20 SKY SIGHTS
(see pages 83–85)

EAST

This dark month we're treated to the brightest supermoon of the year, along with a skyful of glittering stars. You'll notice that the spectacular constellations of winter are gradually drifting westward, and setting earlier, as a result of our annual orbit around the Sun. But, as **Orion** and **Taurus** sink in the west, new constellations appear in the east, including **Virgo** and **Boötes**.

FEBRUARY'S CONSTELLATION

Taurus is very much a second cousin to brilliant **Orion**, but a fascinating constellation nonetheless. It's dominated by **Aldebaran**, the baleful blood-red eye of the celestial bull. Around 65 light years away, and shining with a (slightly variable) magnitude of +0.85, Aldebaran is a red giant star, a tad more massive than the Sun.

The 'head' of the bull is formed by the **Hyades** star cluster. The other famous star cluster in Taurus is the far more glamorous **Pleiades** (the Seven Sisters), whose stars – although further away than the Hyades – are younger and brighter.

Taurus has two 'horns': the star **El Nath** (Arabic for 'the butting one') to the north, and **zeta Tauri** (whose Babylonian name Shurnarkabti-sha-shutu – meaning 'star in the bull towards the south' – is thankfully not generally used!). Above this star is a stellar wreck – literally. In 1054, Chinese astronomers witnessed a brilliant 'new star' appear in this spot, visible in daytime for weeks. It was a supernova – an exploding star in its death throes. And today, we see the still-expanding remains as the **Crab Nebula** (see November's Object). It's visible through a medium-sized telescope.

FEBRUARY'S OBJECT

A pair of objects this month: the beautiful **Double Cluster** in **Perseus**, on the border with **Cassiopeia**. These near-twin star clusters – each covering an area bigger than the Full Moon – are visible to the unaided eye and are a gorgeous sight in binoculars. They're loaded with stunning, young, blue supergiant stars, and lie around 7500 light years away. Officially known as h and chi Persei, each cluster is a mere 12 million years old. Both contain several hundred stars, and are part of what's known as the Perseus OB1 Association – a loose group of bright, hot stars that were born at roughly the same time. Associations and star clusters are important: they allow astronomers to monitor stars that are the same age but have different masses, so providing vital clues to how stars evolve.

OBSERVING TIP

If you want to stargaze at this most glorious time of year, dress up warmly! Lots of layers are better than just a heavy coat, as they trap more air close to your skin, while heavy-soled boots with two pairs of socks stop the frost creeping up your legs. It may sound anorak-ish, but a woolly hat prevents a lot of your body heat escaping through the top of your head. And – alas – no hipflask of whisky. Alcohol constricts the veins, making you feel even colder.

FEBRUARY'S TOPIC: WHAT'S THE DIFFERENCE BETWEEN A STAR AND A PLANET?

Stars shine; planets don't. Simple as that! Hang on, we hear you say: Venus, Jupiter and Mars are brilliant. But that's because they're reflecting light from our local star, the Sun.

Key to the difference is mass. Stars are born in gargantuan clouds of gas and dust – like the Orion Nebula – so there's a lot of cosmic 'stuff' around. When a gas and dust cloud starts to contract under the force of gravity, it breaks up into individual fragments, which continue shrinking separately. As gas gets squeezed, it heats up – like pumping up a flat bicycle tyre. The temperature soars – until its centre reaches ten million degrees.

Then – bang! That's enough to trigger nuclear reactions, which fuse the hydrogen atoms into helium and pour out energy. A star has been born. And starts to shine.

A planet, on the other hand, never achieves the crucial mass to become a star. Planets are made from the leftover debris of starbirth; rocks and gas, and even the most massive planet in our neighbourhood – Jupiter – is only one-thousandth as heavy as the Sun. Although they may have hot cores – enough (as in the case of the Earth) to power plate tectonics – they literally don't have the guts to become luminous.

FEBRUARY'S PICTURE

Every Full Moon is bright, because that's when our companion world is fully illuminated by the Sun. But the Moon moves in an oval (elliptical) orbit, and if it's closest to Earth at the same time, the Full Moon can be up to 14 per cent larger and 30 per cent brighter than the dimmest and most distant Full Moon. Years ago, astrologers (yes, really!) called this phenomenon a 'supermoon' and – because everyone loves a brilliant and romantic Moon – astronomers have jumped onto the same bandwagon. Steve Knight's stunning supermoon image from 2018 shows that the dark maria have variegated colours, reflecting their different chemical composition. This month, there's an even more amazing spectacle on 19 February (see Special Events).

On New Year's Day in 2018, Steve Knight captured the supermoon from his front drive, using a Canon 550D attached to a Skywatcher 8 inch (200 mm) Dobsonian telescope. The image was created from 25 frames (1/4000 sec at ISO 200). Steve stabilised and cropped in PIPP, and processed it in Autostakkert 2 and Faststone Image Viewer, adding colour saturation.

SUNDAY	MONDAY	TUESDAY	WEDNESDAY	THURSDAY	FRIDAY	SATURDAY
					1 Moon near Venus (am)	2 Moon very near Saturn (am)
3	4 9.04 pm New Moon	5	6	7	8	9
10	11	12 10.26 pm First Quarter Moon; Mars near Uranus	13 Moon occults the Hyades	14 Moon occults the Hyades (am)	15	16
17 Moon near Praesepe	18 Moon near Praesepe (am); Venus near Saturn (am)	19 3.53 pm Full Moon near Regulus; supermoon	20	21	22 Moon near Spica	23
24	25	26 11.28 am Last Quarter Moon	27 Moon near Jupiter (am); Mercury E elongation	28 Moon near Jupiter (am)		

Saturn

SPECIAL EVENTS

- **1 February:** around 6.30 am, the crescent Moon lies near Venus, with Jupiter to the upper right.
- **2 February:** the thin crescent Moon rises about 7 am right next to Saturn (in the twilight glow, best seen in binoculars).
- **Night of 13/14 February:** the Moon moves through the Hyades star cluster, near Aldebaran.
- **Night of 17/18 February:** the Moon skims the edge of Praesepe (Chart 2a).
- **18 February:** Venus passes close to Saturn in the dawn twilight (Chart 2b).
- **19 February:** we have a super **supermoon** tonight! A supermoon is a Full Moon

that's nearer than 367,607 km; and this year we have three, in January, February and March. But this month's is the closest, with the Moon lying just 356,846 km away. Enjoy it while it lasts: we'll have to wait almost eight years for a supermoon as good as this.

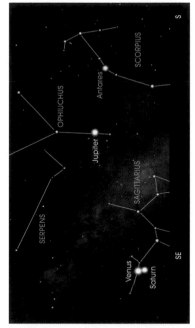

2a 18 February, 4 am. The Moon brushes past the star cluster Praesepe.

2b 18 February, 6 am. Conjunction of Venus and Saturn, with Jupiter and Antares.

- Setting at 9.30 pm, **Mars** shines at magnitude +1.0 as it travels from Pisces to Aries.

- **Neptune** (magnitude +7.9) lurks amid the stars of Aquarius. It's setting about 7 pm, and disappears into the twilight glow by the end of February.

- **Uranus** is on the edge of naked-eye visibility at magnitude +5.8, and is usually difficult to identify among the background stars. But there's a golden opportunity on 12 February, using Mars as your guide. Find the Red Planet (preferably in binoculars), and Uranus is the faint greenish 'star' one degree to the left. This month, Uranus moves into Aries and sets about 9.30 pm.

- In the second half of the month, **Mercury** is visible low down in the west after sunset, reaching greatest eastern elongation on 27 February. It fades from magnitude –1.1 to –0.1 by the end of the month, when it's setting around 7.30 pm.

- It's busy before sunrise! **Venus** rises at 5 am, and hangs like a radiant lantern in the pre-dawn sky, at magnitude –4.2. To the right of the Morning Star you'll find **Jupiter**, shining at magnitude –2.0 in Ophiuchus and rising around 3.30 am.

- At the start of February, **Saturn** lies to the lower left of Venus, way down in the twilight glow in Sagittarius and shining at magnitude +0.6. The two planets are converging, and Saturn lies below Venus on the morning of 18 February, almost a hundred times fainter (Chart 2b).

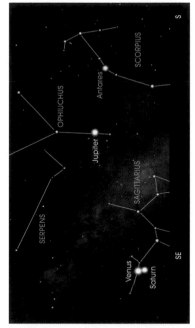

FEBRUARY'S PLANET WATCH

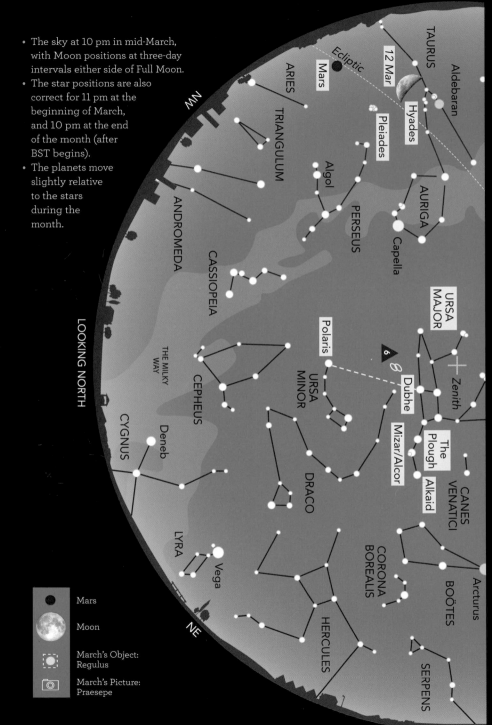

- The sky at 10 pm in mid-March, with Moon positions at three-day intervals either side of Full Moon.
- The star positions are also correct for 11 pm at the beginning of March, and 10 pm at the end of the month (after BST begins).
- The planets move slightly relative to the stars during the month.

WEST

TAURUS

Aldebaran

Ecliptic

Mars

12 Mar

Hyades

ARIES

Pleiades

TRIANGULUM

Algol

AURIGA

NW

PERSEUS

Capella

ANDROMEDA

URSA MAJOR

CASSIOPEIA

Polaris

Zenith

Dubhe

LOOKING NORTH

THE MILKY WAY

CEPHEUS

URSA MINOR

Mizar/Alcor

The Plough

CANES VENATICI

CYGNUS

Deneb

DRACO

Alkaid

LYRA

Vega

CORONA BOREALIS

BOÖTES

Arcturus

NE

HERCULES

SERPENS

Mars

Moon

March's Object: Regulus

March's Picture: Praesepe

EAST

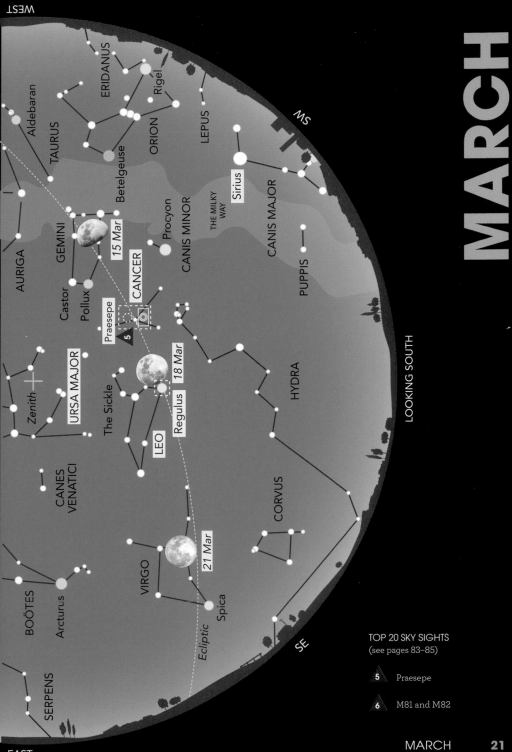

WEST

MARCH

SW

LOOKING SOUTH

SE

EAST

ERIDANUS
Rigel
Aldebaran
TAURUS
ORION
LEPUS
Betelgeuse
15 Mar
GEMINI
Procyon
CANIS MINOR
Sirius
THE MILKY WAY
CANIS MAJOR
AURIGA
Castor
Pollux
CANCER
PUPPIS
Praesepe
5
18 Mar
URSA MAJOR
Zenith
The Sickle
Regulus
HYDRA
LEO
CANES VENATICI
21 Mar
VIRGO
CORVUS
BOÖTES
Arcturus
Spica
Ecliptic
SERPENS

TOP 20 SKY SIGHTS
(see pages 83–85)

5 Praesepe

6 M81 and M82

Spring is here! On 20 March, we celebrate the Equinox, when day becomes longer than the night, while British Summer Time starts on 31 March. Though the nights are shorter, there's still plenty going on in the heavens. Watch **Mars** pass the Seven Sisters – the **Pleiades** star cluster – and the Moon waltzing with Jupiter, Saturn and Venus in the wee small hours.

MARCH'S CONSTELLATION

Ursa Major, the Great Bear, is an internationally favourite constellation. In Britain, its seven brightest stars are called '**The Plough**'. Children today generally haven't seen an old-fashioned horse-drawn plough, and we've found them naming this star-pattern 'the saucepan'. In North America, it's known as 'the Big Dipper'.

The Plough is the first star pattern that most people get to know. It's always on view in the northern hemisphere, and the two end stars of the 'bowl' of the Plough point directly towards the Pole Star, **Polaris**, which always lies due north.

Ursa Major is unusual in a couple of ways. First, it contains a double star that you can split with the naked eye: **Mizar**, the star in the middle of the bear's tail (or the handle of the saucepan) has a fainter companion, **Alcor**.

And – unlike most constellations – the majority of the stars in the Plough lie at the same distance and were born together. Leaving aside **Dubhe** and **Alkaid**, the others are all moving in the same direction (along with brilliant **Sirius**, which is also a member of the group). Over thousands of years, the shape of the Plough will gradually change, as Dubhe and Alkaid go off on their own paths.

MARCH'S OBJECT

Regulus – the 'heart' of **Leo** the Lion – appears to be a bright but run-of–the-mill star. Some 79 light years away, it's young (about a billion years old), 3.8 times heavier than the Sun, and it chucks out 300 times as much energy as our local star. But recent discoveries have revealed it to be a maverick. First, it has at least four companion stars. And it spins in just 16 hours (as compared to roughly a month for our Sun) – meaning that it has a rotational velocity of over a million kilometres per hour. This bizarre behaviour means that its equator bulges like a tangerine. If it were to spin only 10 per cent faster than this, Regulus would tear itself apart!

MARCH'S TOPIC: ORIGIN OF THE ELEMENTS

Where did all the different elements in the Universe – carbon, iron, gold – come from? Back in the 1950s, an ambitious young researcher set up a team to tackle that huge question. Fred Hoyle (or simply Fred, as everyone knew him) chose only the best to work with him: Margaret and Geoffrey Burbidge, a superb observational astronomer and her astrophysicist husband. Then there was Willy Fowler, a Nobel Prize-winning nuclear physicist. And Fred – a brilliant cosmologist. Everyone knew the team by their initials: B²FH.

Astronomers were already aware that the raw material of the Universe was hydrogen, and nuclear reactions in stars like the Sun convert hydrogen to helium. But what happens next? To create the

MARCH'S PICTURE

The **Praesepe** (or Beehive) star cluster lies at the heart of the constellation **Cancer**. Its 1000 stars – two of which are circled by planets – are some 600 million years old, and share a common motion through space with the **Hyades** cluster (which is currently setting in the west). Possibly, both clusters had their origin in the same cloud of dust and gas. In Latin, Praesepe means 'the manger', but the Chinese had a more exotic name for the group: 'the exhalation from piled-up corpses'.

next important element – carbon – Fred suggested a new reaction, that would only occur in heavyweight stars. Off went Willy to his lab, checked the reaction – and Fred's prediction was spot-on.

Now the team worked out how really massive stars could build up successively heavier elements. When each fuel ran out, gravity kicked in and started the next reaction. After carbon came neon, magnesium, silicon – all the elements up to iron.

But... iron has the most stable nucleus of all. Try to fuse iron and the reaction takes in energy. The result: catastrophic collapse. As the star's core shrinks to become an ultra-dense neutron star, its outer layers erupt in a supernova explosion which vomits a cocktail of new elements into space.

Since then, astronomers have found that some of the heaviest elements – such as lead, silver and the gold in your wedding ring – are probably made in the collision of two neutron stars (see May's Topic). But, by proving that the alchemy inside the stars creates the mix of elements that we find around us today, the B²FH paper remains a key scientific breakthrough of the 20th century.

Robin Scagell captured the swarming bees on a Canon A-1 camera with a 135 mm f/2.8 lens. He used Ektachrome film, and the exposure time was 3 minutes.

SUNDAY	MONDAY	TUESDAY	WEDNESDAY	THURSDAY	FRIDAY	SATURDAY
31 British Summer Time begins					1 Moon near Saturn (am)	2 Moon between Venus and Saturn (am)
3 Moon near Venus (am)	4	5	6 4.04 pm New Moon	7	8	9
10	11	12	13 Moon near Aldebaran	14 10.27 am First Quarter Moon	15	16
17	18 Moon near Regulus	19	20 Spring Equinox	21 1.43 am Full Moon	22 Moon near Spica	23
24	25	26 Moon near Antares (am)	27 Moon very near Jupiter (am)	28 4.10 am Last Quarter Moon	29 Moon very near Saturn (am)	30 Mars near Pleiades

The Pleiades

SPECIAL EVENTS

- **1–3 March:** around 5.45 am, watch the crescent Moon pass, first, Saturn and then Venus (Chart 3a).
- **20 March, 9.58 pm:** the Spring Equinox marks the time when the Sun moves up to shine over the northern hemisphere, and days become longer than nights.
- **27 March, 4 am:** the Moon lies right next to brilliant Jupiter.
- **29 March, 4.30 am:** the 'star' just above the crescent Moon is Saturn.
- **30 March:** Mars passes closest to the Pleiades (Seven Sisters) (Chart 3b).
- **31 March, 1.00 am:** British Summer Time starts – don't forget to put your clocks forward (the mnemonic is 'spring forward, fall [autumn] back').

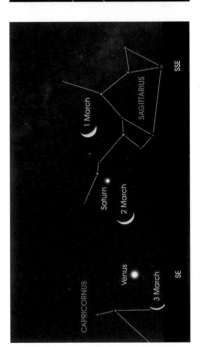

3a 1-3 March, 5.45 am. The crescent Moon passes Saturn and Venus.

3b 30 March, 10 pm. Mars lies close to the Pleiades.

- For a few days only! Look low in the dusk twilight to the west, to spot elusive **Mercury**. At the start of March, it's shining at magnitude +0.1 and setting around 7.30 pm, but the innermost planet is fading and dropping down in sky: a week later it's disappeared from view.

- **Mars** then has the evening sky to itself, until it sets about 11.30 pm. From Aries, the Red Planet speeds upwards into Taurus, shining at magnitude +1.3. On the last couple of nights of March, Mars forms a lovely sight next to the Pleiades (Chart 3b).

- To the lower right of Mars, **Uranus** (magnitude +5.9) lies in Aries, and sets around 9.30 pm.

- The rest of the planetary action takes place in the early hours. Mighty **Jupiter** rises about 2 am in Ophiuchus, a magnificent beacon at magnitude –2.1. The reddish star to its right is Antares. (Look out for the Moon very close by on the morning of 27 March.)

- **Saturn** is next to rise, around 4 am in Sagittarius, and shining at magnitude +0.6. The Moon lies immediately below on the morning of 29 March.

- Last, but very far from least, glorious **Venus** appears low in the morning twilight about 4 am, at a brilliant magnitude –4.1. But the Morning Star is sliding down into the dawn glow, and it will be difficult to spot by the end of March.

- **Neptune** is hidden in the Sun's glare this month.

WEST

- The sky at 11 pm in mid-April, with Moon positions at three-day intervals either side of Full Moon.
- The star positions are also correct for midnight at the beginning of April, and 10 pm at the end of the month.
- The planets move slightly relative to the stars during the month.

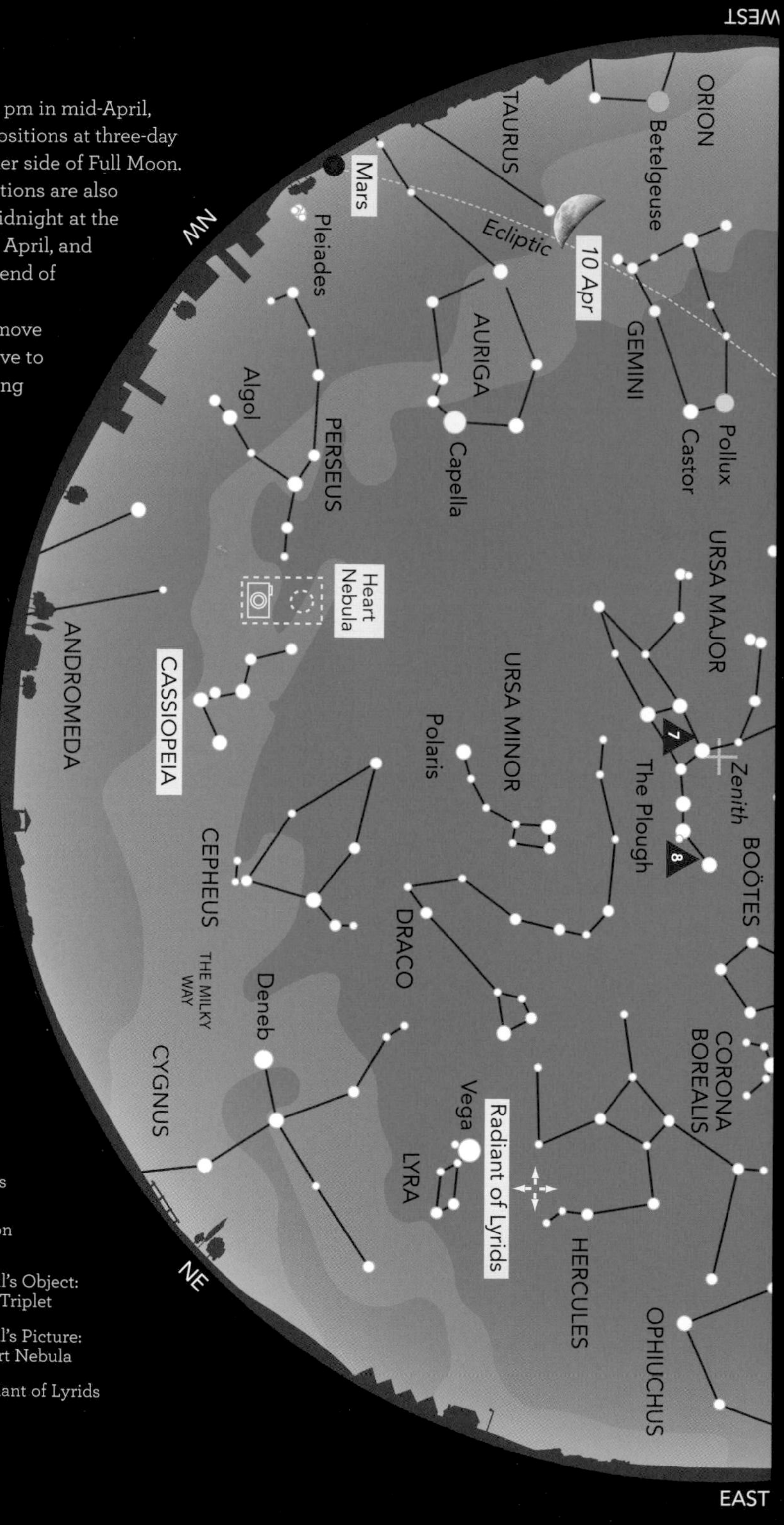

ORION

Betelgeuse

TAURUS

Mars

Pleiades

Ecliptic

10 Apr

GEMINI

Algol

AURIGA

Pollux

Castor

PERSEUS

Capella

URSA MAJOR

Heart Nebula

7

Zenith

BOÖTES

CASSIOPEIA

URSA MINOR

The Plough

8

ANDROMEDA

Polaris

LOOKING NORTH

CEPHEUS

DRACO

CORONA BOREALIS

THE MILKY WAY

Deneb

CYGNUS

Vega

Radiant of Lyrids

LYRA

HERCULES

OPHIUCHUS

NE

Mars

Moon

April's Object: Leo Triplet

April's Picture: Heart Nebula

Radiant of Lyrids

26   APRIL

EAST

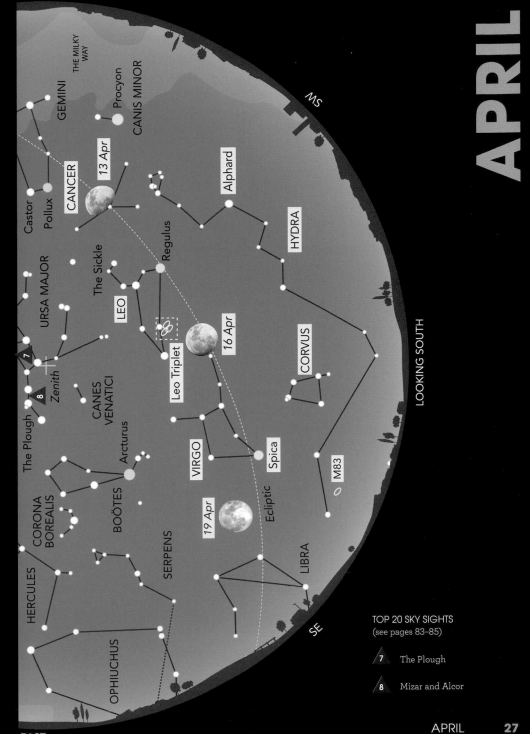

APRIL

WEST

THE MILKY WAY

GEMINI

Procyon

CANIS MINOR

Castor
Pollux

CANCER

13 Apr

Alphard

HYDRA

Regulus

The Sickle

URSA MAJOR

LEO

16 Apr

CORVUS

Leo Triplet

CANES VENATICI

Zenith

7

8

The Plough

Arcturus

VIRGO

Spica

M83

CORONA BOREALIS

BOÖTES

19 Apr

HERCULES

SERPENS

LIBRA

Ecliptic

OPHIUCHUS

SW

LOOKING SOUTH

SE

EAST

TOP 20 SKY SIGHTS
(see pages 83–85)

7   The Plough

8   Mizar and Alcor

APRIL   **27**

The ancient constellations of **Leo** and **Virgo** dominate the springtime skies. Leo does indeed look like a recumbent lion; but it's hard to envisage Virgo as anything other than a vast 'Y' in the sky! There's a parade of planets throughout the night: **Mars** is followed by Jupiter and Saturn, with Venus and Mercury low in the dawn twilight.

APRIL'S CONSTELLATION

Not the most exciting constellation in the heavens, but it's the largest. **Hydra**'s faint stars straggle over a quarter of the sky (100°), along the southern horizon below mighty **Leo**. In legend, it's a fearsome water snake that superhero Hercules had to slay as one of his 12 labours.

But despatching the beast wasn't easy, as it had the irksome habit of growing numerous heads – if one was chopped off, three would grow back! Nothing daunted, Hercules hacked away the extra heads, cauterising the stumps, and buried the last, immortal, head – still hissing – under a stone.

In the heavens, Hydra's head is a pretty grouping of faint stars below **Cancer**. The main star is second-magnitude

Alphard – meaning 'the solitary one'. The snake's body extends under a distinctive quadrilateral of stars depicting **Corvus**, the Crow. And, beneath its elongated tail, use a medium-sized telescope to search out Hydra's hidden gem – the beautiful face-on spiral galaxy **M83**.

APRIL'S OBJECT

Look under the celestial lion's tummy to find the **Leo Triplet**, a trio of spiral galaxies – **M65**, **M66** and **NGC 3628** – that lie around 36 million light years away. They're visible through a small telescope, having magnitudes between +8.9 and +9.5.

These spirals make up a small cluster of galaxies. Like our own Local Group of the Milky Way, Andromeda and the Triangulum Galaxy, the Leo Triplet almost certainly harbours dozens of faint dwarf galaxies.

M65 and NGC 3628 are both seen edge-on, so we can't ascertain much of their structure. But M66 – the brightest of the trio – comes at you full-frontal. It's a glorious galaxy, with curving arms caressing a brilliant nucleus.

NGC 3628 and M66 had a close encounter in the past, leaving NGC 3628 with a dark band of cosmic dust that's a maternity ward for new stars. M66 has a brilliant nucleus – a sign that the entanglement between the two galaxies has already generated starbirth.

OBSERVING TIP

It's best to view your favourite objects when they're well clear of the horizon. If you observe them low down, you're looking through a large thickness of the atmosphere – which is always shifting and turbulent. It's like trying to observe the outside world from the bottom of a swimming pool! This turbulence makes the stars appear to twinkle. Low-down planets also twinkle – although to a lesser extent, because they subtend tiny discs, and aren't so affected.

APRIL'S TOPIC:
THE FERRET OF COMETS

French astronomer Charles Messier (1730–1817) was obsessed with comets. His fascination started when the Great Comet of 1744 put on a spectacular display in the skies over France. It boasted a fan of six tails, and Messier was hooked. He was determined to discover a comet.

But he had a problem. Sweeping the heavens with a refracting telescope boasting a lens 100 mm across, he kept stumbling over fuzzy objects. They *looked* like comets, but they weren't moving. So he and his assistant, Pierre Méchain, decided to log them to avoid confusion.

First on the list was Messier 1 (M1): the Crab Nebula in Taurus, a supernova remnant. M65 and M66 (see this month's Object) are two of the galaxies they bagged. Eventually, the list grew to 110 'Messier Objects' – nebulae, planetary nebulae, galaxies and star clusters.

Messier did get to discover 13 comets, and his obsessive search led to him being called 'the ferret of comets'. But the Messier Catalogue of 110 'deep-sky objects' is

Peter Jenkins captured this image from his home in Kirkby-in-Ashfield, Nottinghamshire, using an Officina Stellare Hiper Apo 115 mm refractor (working at f/5.3) with an Atik 383L+ camera. It comprises 16 x 10-minute exposures each through hydrogen, oxygen and sulphur filters, together with 12 x 30-second exposures each through red, blue and green filters to provide natural star colours. The total exposure time was almost 8½ hours!

his lasting legacy to astronomy. It's used widely even today, and is much loved by amateur astronomers, who race each other to be first to spot the whole lot!

APRIL'S PICTURE

Lying 7500 light years away in the constellation of **Cassiopeia**, the brightest part of the **Heart Nebula** was discovered by William Herschel (of Uranus fame) in 1787. Some 200 light years across, it's an active region of starbirth. The young stars are just 1.5 million years old (as compared to our Sun's 4.6 *billion* years), and some are 50 times more massive than our local star. They live in a loose cluster of stars at the centre of the nebula, and light up their natal gas with their fierce radiation.

SUNDAY	MONDAY	TUESDAY	WEDNESDAY	THURSDAY	FRIDAY	SATURDAY
	1 Moon near Venus	2	3	4	5 9.50 am New Moon	6
7	8	9 Moon near Aldebaran, Mars and the Pleiades	10	11 Mercury W elongation	12 8.06 pm First Quarter Moon	13 Moon occults Praesepe
14 Moon near Regulus	15 Moon near Regulus	16 Venus near Mercury (am)	17 Venus near Mercury (am)	18 Moon near Spica	19 12.12 pm Full Moon	20
21 Moon near Antares	22 Lyrids	23 Lyrids (morning); Moon near Jupiter (am)	24 Moon near Jupiter (am)	25 Moon near Saturn (am)	26 11.18 pm Last Quarter Moon	27
28	29	30				

Lyrid meteor shower

SPECIAL EVENTS

• **9 April:** as the sky darkens, Moon lies just above Aldebaran: to the right, you'll find Mars and then the Pleiades star cluster (Chart 4a).

• **13 April:** between 8.30 pm and midnight, the Moon moves right in front of the Praesepe cluster – best seen in binoculars or a small telescope (Chart 4b).

• **Night of 22/23 April:** maximum of the **Lyrid meteor shower,** which appears to emanate from the constellation of Lyra. The meteors are dusty debris from Comet Thatcher, burning up in the Earth's atmosphere.

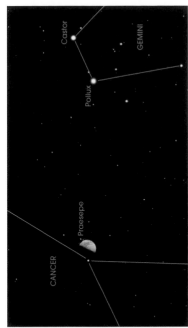

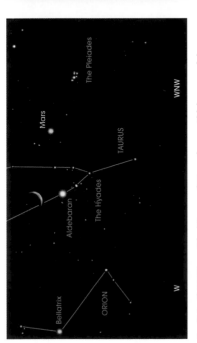

4a 9 April, 10.30 pm. The Moon and Mars close to Aldebaran and the Pleiades.

4b 13 April, 10.30 pm. The Moon occults Praesepe.

Mercury

- **Mars** starts the month next to the Pleiades, the Seven Sisters cluster, and then treks through Taurus, passing brighter Aldebaran in mid-April. As the Earth pulls away from the Red Planet, Mars fades from magnitude +1.4 to +1.6. It sets about midnight.

- Around 1 am, the giant planet **Jupiter** rises in the south-east, blazing brightly at magnitude –2.3 in

Ophiuchus. Fainter **Saturn** (magnitude +0.5) follows on, rising in Sagittarius at about 3 am.

- With a clear eastern horizon, you can spot **Venus** lurking low in the dawn twilight. Shining at magnitude –3.9, the Morning Star peeps above the horizon around 5 am, and is visible for about half an hour before sky fully brightens.

- **Mercury** is at greatest western elongation on 11 April, and we'll see it best in the morning sky after that date. During the second half of the month, try catching the innermost planet down to the lower left of Venus, right on the horizon: binoculars will give your best chance of success. Rising soon after 5 am, Mercury brightens from magnitude +0.3 to –0.3.

- **Uranus** and **Neptune** are too close to the Sun to be seen in April.

APRIL'S PLANET WATCH

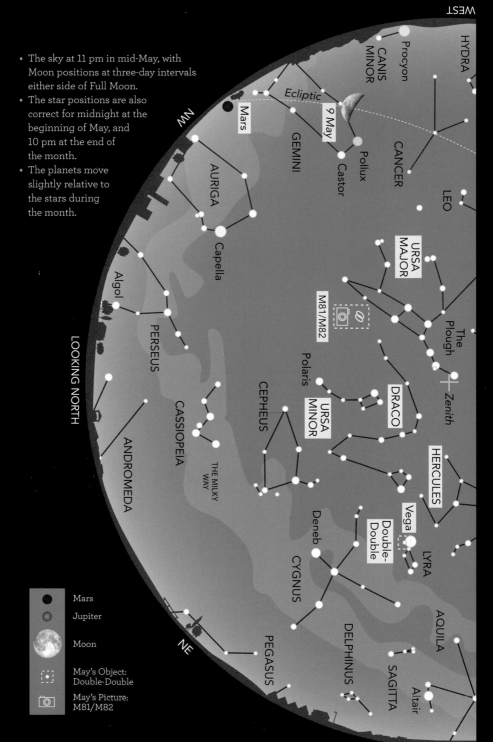

- The sky at 11 pm in mid-May, with Moon positions at three-day intervals either side of Full Moon.
- The star positions are also correct for midnight at the beginning of May, and 10 pm at the end of the month.
- The planets move slightly relative to the stars during the month.

WEST

HYDRA

CANIS MINOR

Procyon

Ecliptic

Mars

9 May

GEMINI

Pollux

Castor

CANCER

LEO

AURIGA

Capella

URSA MAJOR

M81/M82

The Plough

Algol

PERSEUS

Polaris

CEPHEUS

URSA MINOR

DRACO

Zenith

LOOKING NORTH

CASSIOPEIA

THE MILKY WAY

HERCULES

ANDROMEDA

Deneb

Double-Double

Vega

CYGNUS

LYRA

AQUILA

NE

PEGASUS

DELPHINUS

SAGITTA

Altair

Mars

Jupiter

Moon

May's Object: Double-Double

May's Picture: M81/M82

EAST

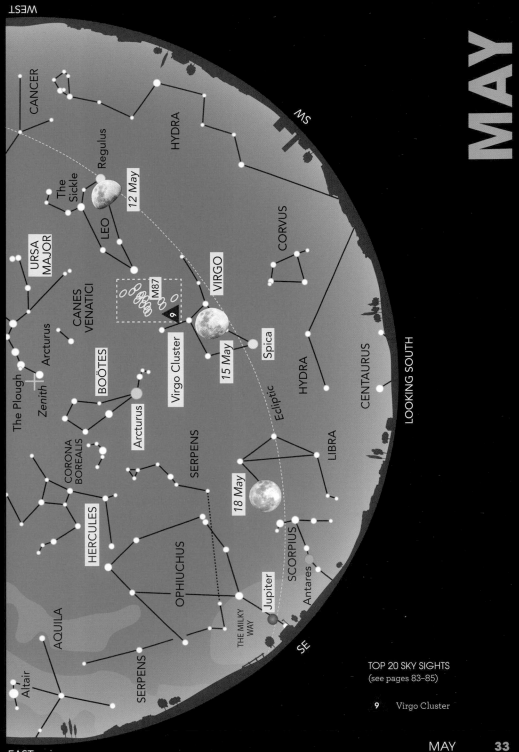

WEST

NW

CANCER

HYDRA

Regulus

The Sickle

12 May

LEO

URSA MAJOR

CANES VENATICI

CORVUS

VIRGO

M87

9

Arcturus

The Plough

Zenith

BOÖTES

Virgo Cluster

15 May

Spica

Ecliptic

HYDRA

CENTAURUS

Arcturus

SERPENS

LOOKING SOUTH

CORONA BOREALIS

18 May

LIBRA

HERCULES

OPHIUCHUS

SCORPIUS

AQUILA

SERPENS

Jupiter

Antares

THE MILKY WAY

SE

Altair

SERPENS

EAST

TOP 20 SKY SIGHTS
(see pages 83–85)

9   Virgo Cluster

Halley's Comet reappears this month! Well – not the actual beast itself, but dirt from its skirt which burns up above our heads as a shower of meteors. Of the stars on view, we have a soft spot for orange-coloured **Arcturus**. The brightest star in **Boötes** (the Herdsman), it shepherds the two bears – **Ursa Major** and **Ursa Minor** – through the heavens, along with an accompanying dragon (**Draco**) and the superhero **Hercules**.

MAY'S CONSTELLATION

The Y-shaped constellation of **Virgo** is the second-largest in the sky. It takes a bit of imagination to see the group of stars as a virtuous maiden holding an ear of corn (the bright star Spica), but this ancient constellation is associated with the time of harvest because the Sun passes through Virgo in the early months of autumn.

Spica is a hot, blue-white star over 12,000 times brighter than the Sun, boasting a temperature of 22,400°C. It has a stellar companion, which lies just 18 million kilometres away from Spica – closer than Mercury orbits the Sun. Both stars inflict a mighty gravitational toll on each other, raising enormous tides and creating two distorted, egg-shaped stars.

OBSERVING TIP

It's always fun to search out the 'faint fuzzies' in the sky – galaxies, star clusters and nebulae. But don't even think of observing dark sky objects around the time of Full Moon – its light will drown them out. You'll have the best views near New Moon: a period astronomers call 'dark of Moon'. When the Moon is bright, though, there's still plenty to see: focus on planets, bright double stars – and, of course, the Moon itself. Check each month's Calendar for the Moon's phases.

The glory of Virgo lies in the 'bowl' of the Y-shape. Scan the upper region with a small telescope – at a low magnification - and you'll find it packed with faint, fuzzy blobs. These are just a few of the 2000 galaxies – star-cities like the Milky Way – that make up the gigantic **Virgo Cluster**. It's centred on the mammoth galaxy **M87**, which boasts a central black hole over five billion times heavier than the Sun.

MAY'S OBJECT

In 1778, the great astronomer William Herschel turned his telescope towards a faint star near brilliant **Vega**, and found it was 'a very curious double-double star. At first sight it appears double... and by attending a little we see that each of the stars is a very delicate double star.'

Although its proper name is epsilon Lyrae, astronomers have nicknamed it ever since the **Double-Double**. Binoculars will show you it's double. Turn a moderately powered telescope (100 mm or larger) in that direction, and you can separate each star into a close double.

The four stars are almost identical quadruplets, about fifth magnitude. Lying 160 light years away, each is a white star almost twice as heavy as our Sun and shining about 15 times brighter. The two stars in the upper pair orbit around each other in 1804 years, while the lower pair complete an orbit in 724 years. But it takes at least 400,000 years for one pair to orbit right around the other pair. Some other faint stars nearby may be held in by their gravity, too, meaning the Double-Double could be a mini-cluster of ten stars.

MAY'S TOPIC: ORIGIN OF LIFE

How common is life in the Universe? Astronomers are starting to find planets like Earth orbiting other stars, where living things could thrive. But has life actually started there?

The best answer comes from studying how life began on Earth. The raw materials of life – water and carbon-rich compounds – not only occurred naturally on Earth, they were delivered in profusion by comets and asteroids impacting our planet. Scientists have several candidates for the natural test tubes where these molecules built up into the building blocks of life: the amino acids that make up proteins and RNA, the precursor of DNA.

Ancient rock pools, alternately filling up and drying up, could have concentrated the primitive compounds into more complex entities. Or they may have assembled at hydrothermal vents, volcanic fissures on the ocean floor where abundant chemical energy could kickstart the first living cells.

Either way, the process seems to have taken just 200–300 million years. Since life arose rapidly on Earth – from common ingredients – the same might apply to the thousands of other worlds that astronomers are discovering in our Galaxy.

MAY'S PICTURE

One of Ursa Major's secrets: the galaxies **M81** and **M82**. These are two members of a galaxy cluster – similar in size to our Local Group – which lies a 'mere' 12 million light years away. Both galaxies are raising cosmic tides on each other. M81 (left) is a serene spiral, exhibiting little signs of ethereal altercation. But for M82 (right), it's a different matter. The interaction between the two galaxies has provoked a violent reaction at the core of this edge-on spiral, leading to a fury of star formation.

Pete Williamson, in Wellington, Shropshire, used an Explore Scientific 102 mm Carbon Fibre apochromatic refractor to ensnare galaxies M81, M82, NGC 2976, NGC 2959 and NGC 2961. He took ten 120-second exposures with a Canon 70D camera

SUNDAY	MONDAY	TUESDAY	WEDNESDAY	THURSDAY	FRIDAY	SATURDAY
			1	2	3	4 11.45 pm New Moon
5	6 Eta Aquarids (am); Moon near Aldebaran	7 Moon near Mars	8	9	10 Moon near Praesepe	11
12 2.12 am First Quarter Moon; Moon near Regulus	13	14	15 Moon near Spica	16 Moon near Spica	17	18 10.11 pm Full Moon
19 Moon near Jupiter and Antares	20 Moon near Jupiter	21 Moon near Jupiter (am)	22	23 Moon near Saturn (am)	24	25
26 5.33 pm Last Quarter Moon	27	28	29	30	31	

Crescent Moon

SPECIAL EVENTS

- **Night of 5/6 May:** shooting stars from the Eta Aquarid meteor shower – tiny pieces shed by Halley's Comet burning up in Earth's atmosphere – fly across the sky tonight. It's an excellent year for these meteors, as the Moon is well out of the way, but they'll only be visible after 3 am.
- **6 May:** low in the evening twilight, the thinnest crescent Moon lies to the right of Aldebaran (best seen in binoculars) (Chart 5a).
- **7 May:** the reddish 'star' above the crescent Moon is the planet Mars (Chart 5a).
- **Night of 19/20 May:** the Moon forms a triangle with Jupiter (lower left) and Antares (lower right) (Chart 5b).
- **Night of 20/21 May:** the Moon lies just to the left of brilliant planet Jupiter (Chart 5b).

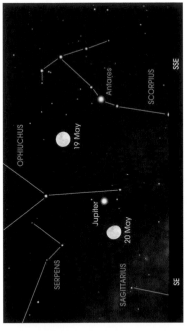

5a 6, 7 May, 9.15 pm. The crescent Moon with Aldebaran and Mars.

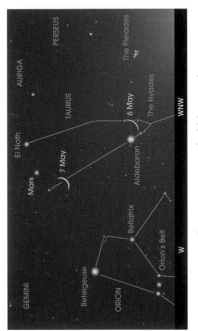

5b 19, 20 May, 11.50 pm. The Moon with Jupiter and Antares.

Mars

• In the evening sky, **Mars** still holds sway as it gradually sinks into the twilight sky, setting around midnight. The Red Planet shines at magnitude +1.7, and moves from Taurus to Gemini.

• During the last few days of May, Mars has serious competition as **Mercury** roars up above the horizon in the north-west. At magnitude –1.4, the innermost planet is 20 times brighter than Mars, and by the end of the month it's setting well after 10 pm.

• **Jupiter** is now rising about 11 pm. The giant planet shines at magnitude –2.5 in Ophiuchus. Well to its left, you'll find **Saturn** some 15 times fainter (magnitude +0.4), rising in Sagittarius around 1 am.

• Outermost planet **Neptune** rises in the morning sky

about 3 am, at a dim magnitude +7.9 in Aquarius.

• Finally, you can catch brilliant **Venus** (magnitude –3.9) very low in the east, rising about 4.30 am – just before the Sun comes up.

• **Uranus** is lost in the Sun's glare this month.

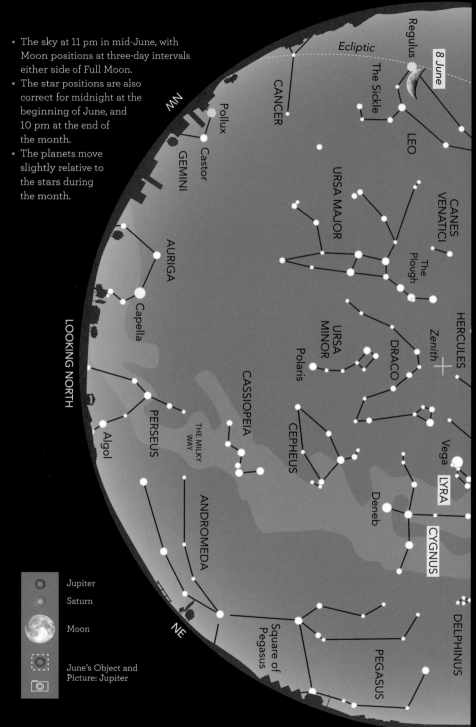

- The sky at 11 pm in mid-June, with Moon positions at three-day intervals either side of Full Moon.
- The star positions are also correct for midnight at the beginning of June, and 10 pm at the end of the month.
- The planets move slightly relative to the stars during the month.

WEST

Regulus

Ecliptic

8 June

The Sickle

CANCER

LEO

NW

Pollux

Castor

GEMINI

URSA MAJOR

CANES VENATICI

The Plough

AURIGA

Capella

HERCULES

Zenith

URSA MINOR

DRACO

Polaris

LOOKING NORTH

CASSIOPEIA

CEPHEUS

PERSEUS

THE MILKY WAY

Vega

Algol

LYRA

ANDROMEDA

Deneb

CYGNUS

NE

Square of Pegasus

DELPHINUS

PEGASUS

EAST

	Jupiter
	Saturn
	Moon
	June's Object and Picture: Jupiter

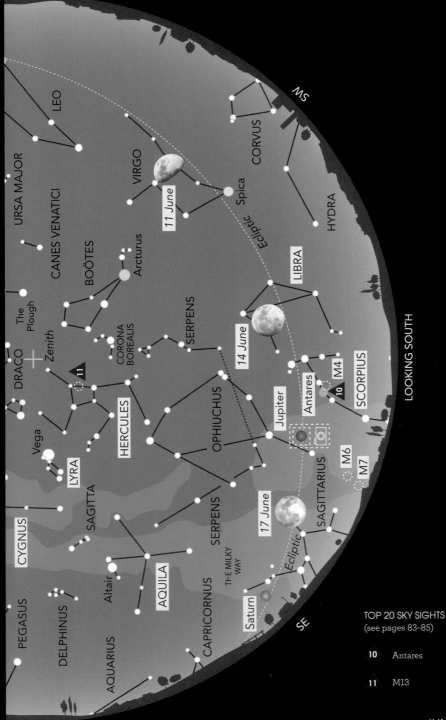

WEST

LEO

URSA MAJOR

CANES VENATICI

BOÖTES

The
Plough

Zenith

DRACO

Arcturus

CORONA
BOREALIS

HERCULES

Vega

LYRA

SAGITTA

CYGNUS

PEGASUS

DELPHINUS

AQUILA

Altair

AQUARIUS

CAPRICORNUS

EAST

11

VIRGO

11 June

Spica

Ecliptic

SERPENS

SERPENS

14 June

OPHIUCHUS

Jupiter

Antares

10

M4

SCORPIUS

M6

M7

SAGITTARIUS

17 June

THE MILKY
WAY

Ecliptic

Saturn

SE

CORVUS

HYDRA

LIBRA

SW

LOOKING SOUTH

TOP 20 SKY SIGHTS
(see pages 83–85)

10   Antares

11   M13

With the sky never quite getting dark – especially in the north of the country – it's not the greatest month for spotting faint stars. But take advantage of the soft, warm weather to acquaint yourself with the lovely summer constellations of **Hercules**, **Scorpius**, **Lyra**, **Cygnus** and **Aquila**. Not to mention brilliant **Jupiter,** putting on its best show of the year.

JUNE'S CONSTELLATION

Down in the deep south of the sky lies a baleful red star. **Antares** – 'the rival of Mars' – surpasses in ruddiness even the famed Red Planet. To the ancient Greeks, Antares marked the heart of **Scorpius**, the celestial scorpion. They intimately linked this summer constellation with the winter-star pattern Orion, who was killed by a mighty scorpion. The gods placed both in the sky, but at opposite sides, so Orion sets as Scorpius rises.

Unusually, Scorpius resembles its terrestrial namesake. A line of stars to the top right of Antares marks the scorpion's forelimbs. Originally, the stars we now call **Libra** (the Scales) were its claws. Below Antares, the scorpion's body stretches down into a fine curved tail (below the horizon on the chart), and deadly sting. Alas – these aren't visible from the latitude of the UK: an excuse for a Mediterranean holiday!

Scorpius is a treasure trove of astronomical goodies. Several lovely double stars include Antares: its faint companion looks greenish in contrast to Antares's strong red hue. Binoculars reveal the fuzzy patch of **M4**, a globular cluster made of tens of thousands of stars, some 7,200 light years away. Above the 'sting' lie two fine star clusters – **M6** and **M7** – visible to the naked eye when they're well above the horizon: a telescope reveals their stars clearly.

JUNE'S OBJECT

At 143,000 kilometres in diameter, **Jupiter** is the biggest world in our Solar System. It could contain 1300 Earths – and the cloudy gas giant is very efficient at reflecting sunlight. Jupiter now is shining at a dazzling magnitude –2.6, and it's a fantastic target for stargazers, whether you're using your unaided eyes, binoculars or a small telescope.

Despite its size, Jupiter spins faster than any other planet, in 9 hours 55 minutes. As a result, its equator bulges outwards – through a small telescope, it looks like a squashed orange crossed with an old-fashioned humbug. The stripes are cloud belts of ammonia and methane stretched out by the planet's dizzy spin.

OBSERVING TIP

Don't think that you need a telescope to bring the heavens closer. Binoculars are excellent – and you can fling them into the back of the car at the last minute. For astronomy, buy binoculars with large lenses coupled with a modest magnification. An ideal size is 7 x 50, meaning that the magnification is seven times, and that the diameter of the lenses is 50 millimetres. They have good light grasp, and the low magnification means that they don't exaggerate the wobbles of your arms too much. Even so, it's best to rest your binoculars on a wall or a fence to steady the image.

Jupiter commands a family of about 70 moons. The four biggest are visible in good binoculars, and even – for the really sharp-sighted – to the unaided eye. In the 2020s, the European Space Agency will send the JUICE spaceprobe to explore the icy moons Ganymede, Callisto and Europa. The last will also be the target of NASA's Europa Clipper, a spacecraft that will check out a suspected giant ocean under the frozen surface – where, possibly, primitive life may flourish.

Damian Peach composed this image from around 15,000 frames taken through IR, G, B filters, with an ASI 174MM camera on the 1 m telescope at the Pic du Midi in France. He used an IR (infrared) filter rather than a R (red) filter because the IR image is generally the sharpest.

JUNE'S TOPIC: GRAVITATIONAL WAVES

A century ago, Albert Einstein showed that space behaves like an invisible rubber sheet filling the entire Cosmos. Something heavy, like a star, makes a dent in the sheet, and an orbiting planet feels that dent as the force of the star's gravity. Einstein also realised that two heavy objects orbiting rapidly would stir up waves in the 'rubber sheet'. These gravitational waves would spread out across the Universe, like ripples in a pond when you stir it with a stick.

Scientists have built sensitive instruments to detect gravitational waves, in the United States and in Italy. In 2015, they picked up the first tiny shudder from across the Universe – ripples from two massive black holes spinning around each other and then coalescing. Two years later, they detected gravitational waves from a pair of neutron stars in a similar dance of death. Alerted by the discovery, other astronomers imaged the catastrophic merger of the two stars, and found a vast amount of newly created matter surging outwards – including ten times the Earth's weight in gold!

JUNE'S PICTURE

This stunning portrait of **Jupiter** also shows its biggest moon, Ganymede. Jupiter's atmosphere is ever changing. And that's also true of the Great Red Spot (lower right), which has been shrinking over the past decades. Once three times wider than the Earth, it's now only the diameter of our planet.

SUNDAY	MONDAY	TUESDAY	WEDNESDAY	THURSDAY	FRIDAY	SATURDAY
30						1
2	3 11.02 am New Moon	4	5 Moon near Mars	6	7	8 Moon near Regulus
9	10 6.59 am First Quarter Moon; Jupiter opposition	11	12 Moon near Spica	13	14	15 Moon near Jupiter and Antares
16 Moon very near Jupiter	17 9.31 am Full Moon	18 Mercury very close to Mars; Moon near Saturn	19 Moon near Saturn (am)	20	21 Summer Solstice	22
23 Mercury E elongation	24	25 10.46 am Last Quarter Moon	26	27	28	29

SPECIAL EVENTS

- **10 June:** Jupiter is opposite to the Sun in the sky. Because the planets' orbits are not quite circular, we are closest to the giant planet on 12 June, at 641 million km. With binoculars or a small telescope, check out Jupiter's four biggest moons.

- **15 June:** the almost-Full Moon lies above Antares, with Jupiter to the left (Chart 6a).

- **16 June:** the brilliant 'star' just to the right of the Moon is giant planet Jupiter, with Antares farther off in the same direction; to the Moon's left, you'll find Saturn (Chart 6a).

- **18 June, 10.30 pm:** look very low to the north-west to spot Mercury almost on the horizon, with twin stars Castor and Pollux above. Whip out binoculars for a better view, and you'll also spot Mars just below – and very close to – Mercury (just 14 arcminutes away) (Chart 6b).

- **Night of 18/19 June:** the Moon passes below Saturn (Chart 6a).

- **21 June, 4.54 pm:** Summer Solstice. The Sun reaches its most northerly point in the sky, so today is Midsummer's Day, with the longest period of daylight and the shortest night.

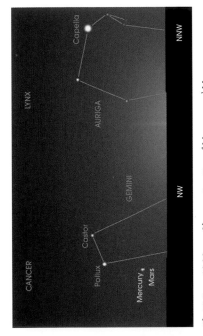

6a 15-18 June, 11.50 pm. The Moon with Jupiter, Antares and Saturn.

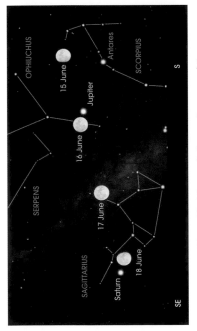

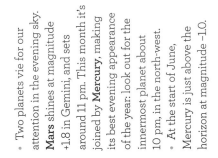

6b 18 June, 10.30 pm. Close conjunction of Mercury and Mars.

- Two planets vie for our attention in the evening sky. **Mars** shines at magnitude +1.8 in Gemini, and sets around 11 pm. This month it's joined by **Mercury**, making its best evening appearance of the year: look out for the innermost planet about 10 pm, in the north-west.
- At the start of June, Mercury is just above the horizon at magnitude –1.0. The tiny world steams upwards towards fainter Mars, fading as it goes. Mercury is at magnitude +0.2 – three times brighter than Mars – when the planets pass very close on 18 June (see Special Events). Thereafter, Mercury loops downward, fading to magnitude +1.1 by the end of June.
- **Jupiter** is now visible all night long, at its closest to

Earth (see Special Events). The giant planet lies in Ophiuchus, and it's by far the brightest object in the night sky (bar the Moon), at magnitude –2.6.
- Rising around 11 pm, fainter **Saturn** is shining at magnitude +0.3 in Sagittarius. It's followed by **Neptune** – so dim, at magnitude +7.9 that you need binoculars or a telescope to see it –

lying in Aquarius and rising about 1 am.
- Next up above the horizon, around 2.30 am, is Neptune's twin planet, **Uranus**, lying in Aries and glowing at magnitude +5.9.
- And, finally, there's **Venus**. Rising just before the Sun, at 4 am, you may catch this brilliant world (magnitude –3.9) hugging the horizon to the north-east all month.

- The sky at 11 pm in mid-July, with Moon positions at three-day intervals either side of Full Moon.
- The star positions are also correct for midnight at the beginning of July, and 10 pm at the end of the month.
- The planets move slightly relative to the stars during the month.

7 July

VIRGO

LEO

The Sickle

CANES VENATICI

BOÖTES

The Plough

NW

URSA MAJOR

HERCULES

DRACO

AURIGA

Polaris

URSA MINOR

Zenith

CYGNUS

Capella

CASSIOPEIA

CEPHEUS

Deneb

THE MILKY WAY

North American Nebula

LOOKING NORTH

PERSEUS

Algol

PEGASUS

Square of Pegasus

TRIANGULUM

ANDROMEDA

NE

PISCES

Jupiter

Saturn

Neptune

Moon

July's Object: Lagoon and Trifid Nebulae

EAST

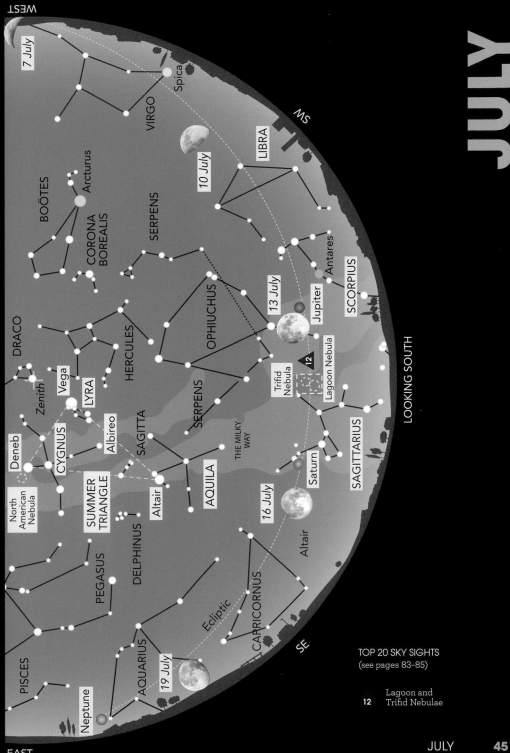

JULY

7 July

Spica

VIRGO

MS

10 July

LIBRA

BOÖTES

Arcturus

CORONA
BOREALIS

SERPENS

13 July

Antares

Jupiter

SCORPIUS

DRACO

HERCULES

OPHIUCHUS

Zenith

Vega

LYRA

12

Trifid
Nebula

Lagoon Nebula

LOOKING SOUTH

Deneb

CYGNUS

Albireo

SAGITTA

SERPENS

THE MILKY
WAY

Saturn

SAGITTARIUS

North
American
Nebula

SUMMER
TRIANGLE

Altair

AQUILA

16 July

Altair

PEGASUS

DELPHINUS

CARRICORNUS

SE

TOP 20 SKY SIGHTS
(see pages 83–85)

PISCES

AQUARIUS

19 July

Ecliptic

Neptune

12   Lagoon and
      Trifid Nebulae

We've a partial eclipse of the Moon to savour this month, though the supreme sight of a total solar eclipse is reserved for those trekking to South America. **Saturn** has a starring role low down in Sagittarius, surrounded by the jewels of the Milky Way. We're also treated to lovely stars such as **Vega, Deneb** and **Altair**, which comprise the **Summer Triangle**.

JULY'S CONSTELLATION

The flying swan – **Cygnus** – is one of our most cherished constellations. The celestial bird actually looks like its namesake, with outspread wings and an elongated neck.

Swan legends abound. One of the most popular is that Zeus – disguised as a swan – seduced Leda, the wife of King Tyndareus of Sparta. As a result, the unfortunate woman gave birth to twins, one immortal and one mortal: they appear in the sky as Pollux and Castor, the heavenly twins in the constellation of Gemini.

Deneb forms the swan's tail. It's the furthest away of the first-magnitude stars, but its distance is hard to pin down. Estimates range from 1500 to 2600 light years: which means that the star shines anything between 50,000 and 200,000 times brighter than the Sun.

OBSERVING TIP

This is the month when you really need a good, unobstructed horizon to the south, for the best views of the iconic summer constellations of Scorpius and Sagittarius. They never rise high in temperate latitudes, so make the best of a southerly view – especially over the sea – if you're away on holiday. A good southern horizon is also best for views of the planets, because they rise highest when they're in the south.

The swan's head is marked by what's probably the most beautiful double star in the sky: **Albireo**. You need a telescope to gaze upon this gold-and-sapphire jewel.

Another Cygnus gem is the **North America Nebula**. Looking uncannily like its terrestrial namesake, this glowing cloud of gas is bigger than the Full Moon, but it's so faint that you'll need binoculars to pick it out.

The Milky Way meanders through Cygnus, split by the dark silhouette of the Great Rift. Edge-on, this band of star soot is riddled with star clusters and nebulae. It's fantastic to sweep Cygnus with binoculars or a small telescope.

JULY'S OBJECT

Sagittarius – alas, too far south to be a sensational sight from Britain – is home to a pair of glorious nebulae. These gaseous crucibles of starbirth are among the most heavenly sights in the sky (sorry about the pun). The **Lagoon Nebula** is an oasis of calm: this gentle womb of burgeoning stars is home to many 'Bok Globules' – small black clouds hatching baby suns.

Its companion, the **Trifid Nebula**, looks dramatically different. It's dissected by bands of dark cosmic dust – interstellar soot – which is poised to collapse and create fledgling stars. NASA's Spitzer orbiting telescope, which looks at the Universe in infrared (heat rays), has discovered 120 newborn stars in the nebula. The Trifid lies just over 5000 light years away.

JULY'S TOPIC:
THE SOLAR WIND

Amazingly, the Earth orbits the Sun *inside* its atmosphere! Streaming out of our local star is the solar wind, a gale of protons and electrons from the Sun's corona (see this month's Picture).

Close to the Sun, the wind is supersonic, travelling at speeds up to 750 kilometres per second. But where does it come from? Not from the heat of the Sun's surface: that's for sure. It was first predicted in the mid-19th century by British astronomer Richard Carrington after a magnetic flare on the Sun's surface caused a coronal mass ejection in the Sun's atmosphere. Charged particles from the explosion hurtled towards Earth, resulting in a solar storm. Astronomers agreed that the cause was the Sun's erratic magnetic activity.

The solar wind drives the 'space weather' in our Solar System. It sweeps back the magnetised gas tails of comets, and it's capable of partially stripping the atmospheres of small planets, like Mars.

The Earth is protected from the solar wind by its own protective magnetic field. But it's not always foolproof. Satellites orbiting high above our planet have been knocked out by solar storms, and they're also a hazard for astronauts.

The Sun's charged particles are also attracted to Earth's magnetic poles, lighting up the atmosphere in spectacular displays of the aurorae. But its effects are not always benign. In 1989, a powerful solar storm hit Canada, wiping out the powerlines in Quebec for hours. It also disrupted the Toronto Stock Exchange. All trading stopped when the storm made the computers crash. Now *that's* a disaster!

JULY'S PICTURE

We've chosen a stunning image of a past solar eclipse to give an idea of the treat in store for parts of South America this month. Eclipses of the Sun are the result of sheer coincidence. The Sun is 400 times wider than the Moon; but it's also 400 times further away. When the paths of the two bodies cross, the Moon can overlap the Sun exactly, revealing the delicate tendrils of our star's outer atmosphere – the corona.

Alan Dowdell took this image of the eclipsed Sun on 21 August 2017 at Riverton, Wyoming. He used a Canon 600D DSLR camera attached to an Explore Scientific 66 mm, 500 mm focal length refracting telescope. The exposure was 1/40 second at ISO 400.

SUNDAY	MONDAY	TUESDAY	WEDNESDAY	THURSDAY	FRIDAY	SATURDAY
	1	2 8.16 pm New Moon; solar eclipse	3	4 Earth at aphelion; Moon near Mercury, Mars and Praesepe	5 Moon near Regulus	6
7	8	9 11.55 am First Quarter Moon near Spica; Saturn opposition	10	11	12 Moon near Jupiter and Antares	13 Moon very near Jupiter
14	15 Moon near Saturn	16 10.38 pm Full Moon; lunar eclipse	17	18	19	20
21	22	23	24	25 2.18 am Last Quarter Moon	26	27
28 Moon near Aldebaran (am)	29	30	31			

SPECIAL EVENTS

• **2 July:** The Sun is totally eclipsed from a narrow track across the southern Pacific Ocean, Chile and Argentina. South America will witness a partial eclipse, but nothing is visible from the UK.

• **4 July, 9.45 pm:** look low in the twilight to the north-west (preferably with binoculars) to see the thinnest crescent Moon with Mercury and Mars (Chart 7a).

• **4 July, 11.11 pm:** the Earth is furthest from the Sun (aphelion), at 152 million km.

• **9 July:** Saturn is opposite to the Sun in the sky and closest to Earth, 1351 million km away.

• **12 July:** the Moon forms a triangle with Jupiter (left) and Antares (lower left).

• **13 July:** Jupiter lies just below the Moon, with Antares to the lower right and Saturn well to the left (Chart 7b).

• **16 July:** Look out for a partial lunar eclipse, visible in Europe, Africa, South America and western Asia. From Britain, we'll see the Full Moon rising around 9 pm already partly obscured by the Earth's shadow; maximum eclipse occurs at 10.32 pm, when the Moon is 65 per cent obscured, and it emerges from eclipse at midnight (Chart 7b).

• **28 July, 3 am:** the crescent Moon lies just to the left of Aldebaran and the Hyades, with the Pleiades above.

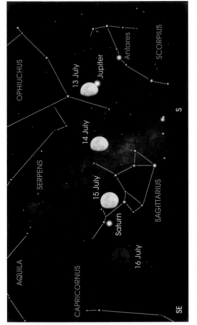

7a 4 July, 9.45 pm. Conjunction of Mercury and Mars.

7b 13-16 July, 11 pm. The Moon with Jupiter, Antares and Saturn; total lunar eclipse.

- At the beginning of the month, you can just catch **Mercury** and **Mars** very low in the north-west after sunset, in Gemini (Chart 7a), forming a line of four roughly equal objects with the constellation's main stars: from left to right they are Mercury (magnitude +1.8), Mars (+1.2), Pollux (+1.2) and Castor (+1.6). But they have disappeared into the twilight glow by mid-July.

- The main action is taking place in the southern sky, where **Jupiter** is blazing brilliantly at magnitude −2.5 in Ophiuchus, until it sets around 2.30 am. The star to its lower right is Antares, in Scorpius.

- To the left you'll find **Saturn**, at its brightest and closest to Earth on 9 July (see Special Events); even then, it's ten times fainter

than Jupiter at magnitude +0.1. Visible all night long, Saturn lies in Sagittarius. Grab a telescope to view its magnificent rings and giant moon Titan.

- Next comes **Neptune**, rising about 11 pm in Aquarius. You'll need binoculars or a telescope to spot this faint world. a mere magnitude +7.8. **Uranus** appears above the

horizon around 0.30 am, at magnitude +5.8 in Aries.

- Just before dawn, you may just catch brilliant **Venus** very low on the north-eastern horizon. Shining at magnitude −3.9, it rises just after 4 am.

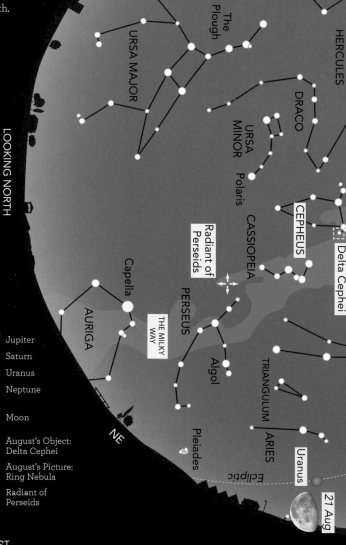

WEST

- The sky at 11 pm in mid-August, with Moon positions at three-day intervals either side of Full Moon.
- The star positions are also correct for midnight at the beginning of August, and 10 pm at the end of the month.
- The planets move slightly relative to the stars during the month.

NW

LOOKING NORTH

Arcturus

CORONA BOREALIS

BOÖTES

CANES VENATICI

HERCULES

The Plough

URSA MAJOR

DRACO

URSA MINOR

Polaris

Zenith

CYGNUS

Deneb

CEPHEUS

Radiant of Perseids

CASSIOPEIA

Delta Cephei

PEGASUS

Capella

PERSEUS

THE MILKY WAY

Algol

TRIANGULUM

ANDROMEDA

AURIGA

ARIES

Uranus

Ecliptic

PISCES

NE

Pleiades

21 Aug

Jupiter
Saturn
Uranus
Neptune

Moon

August's Object:
Delta Cephei

August's Picture:
Ring Nebula

Radiant of
Perseids

EAST

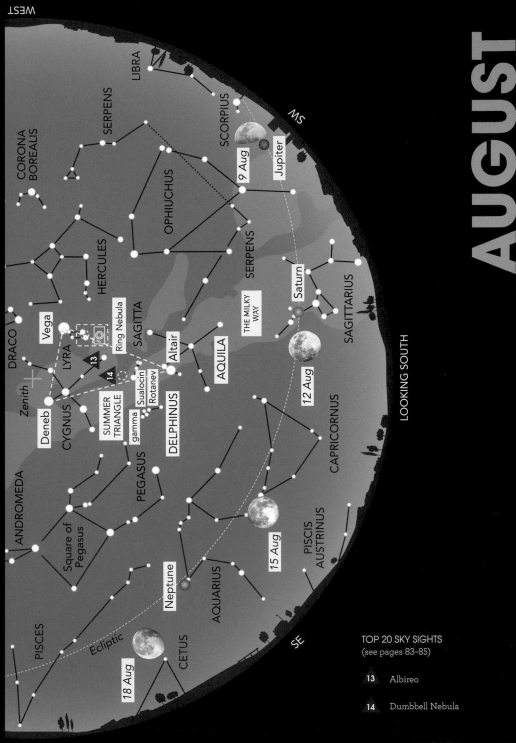

WEST

LIBRA

SERPENS

CORONA
BOREALIS

SCORPIUS

9 Aug

MS

Jupiter

DRACO

OPHIUCHUS

HERCULES

SERPENS

Saturn

SAGITTARIUS

Vega

SAGITTA

Ring Nebula

THE MILKY
WAY

LYRA

13

Zenith

14

Altair

AQUILA

Sualocin
Rotanev

12 Aug

Deneb

SUMMER
TRIANGLE

gamma

DELPHINUS

CYGNUS

LOOKING SOUTH

ANDROMEDA

PEGASUS

CAPRICORNUS

Square of
Pegasus

15 Aug

PISCIS
AUSTRINUS

Neptune

AQUARIUS

PISCES

Ecliptic

SE

18 Aug

CETUS

EAST

The giants of the Solar System, **Jupiter** and **Saturn,** lie low in the south this month: above them, the **Milky Way** arches right up over the heavens. Also watch out for shooting stars from the annual **Perseid** meteor shower mid-month, though this year the display is spoilt by bright moonlight.

AUGUST'S CONSTELLATION

It may be small, but it's perfectly formed. **Delphinus,** the celestial dolphin, is outlined by four stars making a lopsided rectangle, with an extra star forming his tail. To fish out this constellation, first locate the **Summer Triangle** of the bright stars **Vega, Deneb** and **Altair,** then look to the upper left of Altair.

This constellation immortalises humanity's long relationship with the most intelligent marine life on our planet. According to one myth, the dolphin acted as go-between when the sea god Poseidon (Neptune) was courting his wife, the sea nymph Amphitrite. In another story, the dolphin rescued the musician Arion when he was thrown overboard by sailors intent on stealing his wealth.

The two stars to the right of the rectangle are called **Sualocin** and **Rotanev.** These strange-looking names represent a bit of self-promotion by a 19th-century Italian astronomer, Niccolo Cacciatore. In Latin, his name becomes Nicolaus Venator: try spelling this backwards!

The top-left star of Delphinus, **gamma Delphini,** is a lovely double star when you observe it with a reasonable telescope.

AUGUST'S OBJECT

At first glance, the star **Delta Cephei** – in the constellation representing King **Cepheus** – doesn't seem to merit any special attention. It's a yellowish star of magnitude +4 – easily visible to the naked eye, but not prominent. A telescope reveals a companion star. But this star holds the key to measuring the size of the Universe.

Check the brightness of this star carefully over days and weeks, and you'll see that its brightness changes regularly, from +3.5 (brightest) to +4.4 (faintest), every 5 days 9 hours. It's a result of the star literally swelling and shrinking in size, from 32 to 35 times the Sun's diameter.

Stars like this – Cepheid variables – have a link between their period of variation and their intrinsic luminosity. Astronomers can deduce a Cepheid's true luminosity by observing the star's period, and by comparing this to the brightness it appears in the sky, they can work out the star's distance. With the Hubble Space Telescope, astronomers have now measured the most distant Cepheids in the galaxy NGC 4603, which lies 107 million light years away.

AUGUST'S TOPIC:
NAMING OF EXOPLANETS

With around 4000 planets now known to orbit other stars, astronomers have a slight naming problem on their hands! No worries. Turn to the official body that 'governs' astronomy – the International Astronomical Union (IAU) – and bring in their formidable Nomenclature Committee.

The process started quite tamely. A new planet was named after its parent star, and allotted the name 'b' (the star itself is always 'a'). Subsequent planets would be labelled c, d, e...

But private companies got in on the act, asking the public to pay to name an exoplanet. Not to be daunted, the IAU fought back. In 2014, they invited the public to name exoplanets officially, and put their names to a vote.

Galileo, Brahe, Hypatia, Lipperhey, Quijote – and Poltergeist (named because it circles a pulsar, a dead star) – now grace our heavens: official.

In February 2017, NASA's Spitzer space telescope discovered seven planets circling a dim

Ed Cloutman of Swansea first imaged the Ring Nebula through an Ikharos 250 mm Ritchey-Chretien reflector telescope and a QSI 683WS camera. He augmented the picture with further exposures using a Takahashi 130 mm refractor, bringing the total exposure time to 2 hours.

red dwarf star, TRAPPIST 1. The space agency – in a fit of fun – asked the public to name them. Thank goodness they're not official. Otherwise, we'd have Harry Potter and Donald Trump up there!

AUGUST'S PICTURE

The **Ring Nebula** looks like a celestial smoke-ring. William Herschel – who discovered Uranus – found many similar puzzling objects. They looked very similar to the planet he had stumbled over. So he called them 'planetary nebulae'.

But the Ring Nebula is a dying star. Its core has run out of nuclear fuel, and the unstable star has puffed off its outer layers into space. Eventually, these layers will disperse, leaving the core exposed as a cooling white dwarf star – which will later become a black, celestial cinder.

Lying between the two lowest stars of the tiny constellation **Lyra**, the Ring Nebula is faint (magnitude +9) and only 1 arcminute across. It's best to have a 200 mm telescope or larger to observe the planetary nebula well.

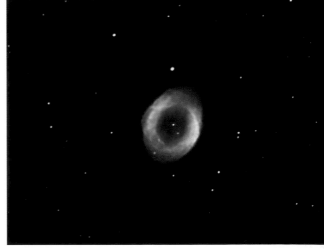

SUNDAY	MONDAY	TUESDAY	WEDNESDAY	THURSDAY	FRIDAY	SATURDAY
				1 4.12 am New Moon	2	3
4	5 Moon near Spica	6 Moon near Spica	7 6.31 pm First Quarter Moon	8	9 Mercury W elongation; Moon near Jupiter and Antares	10
11 Moon near Saturn	12 Jupiter nearest to Antares; Perseids	13 Perseids (am)	14	15 1.29 pm Full Moon	16	17
18	19	20	21	22	23 3.56 pm Last Quarter Moon Moon occults the Hyades	24 Moon occults the Hyades (am)
25	26	27	28	29	30 11.37 am New Moon	31

Hyades

SPECIAL EVENTS

• **9 August:** the Moon lies above Jupiter, with Antares below and Saturn well to the left (Chart 8a).

• **Night of 12/13 August:** maximum of the **Perseid meteor shower**. The Earth runs into debris from Comet Swift-Tuttle, which streams into the atmosphere at speeds of 210,000 km/h and burns up 60 km above the Earth's surface. This year, unfortunately, all but the brightest Perseids will be washed out by moonlight.

• **Night of 23/24 August:** when the crescent Moon rises, around 1 am, you'll find it well below the Pleiades, and right on the edge of the Hyades star cluster (near the bright star Aldebaran). Watch over the next few hours (preferably with binoculars or a small telescope), to see the Moon hide several of the Hyades stars in turn (Chart 8b).

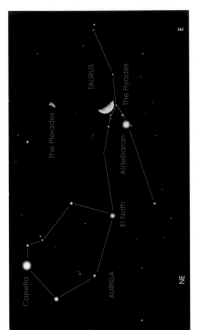

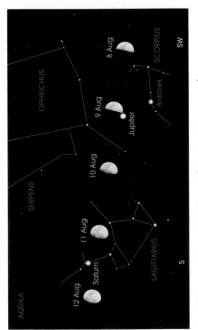

8a 8-12 August, 11 pm. The Moon with Jupiter, Antares and Saturn.

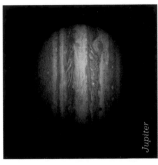

8b 24 August, 1 am. The Moon occults the Hyades.

• Magnificent **Jupiter** rules the evening sky, appearing low in the southwest in Ophiuchus, above the red giant star Antares. At magnitude −2.3, the giant planet is the brightest object in the night sky (after the Moon). It sets about midnight.

• **Saturn** lies higher in the sky, to the left, in Sagittarius. Setting around 2.30 am, the ringworld shines at magnitude +0.2.

Jupiter

• **Neptune** lurks in Aquarius, at a dim magnitude +7.8, and is visible all night long. Its planetary twin, **Uranus** (magnitude +5.8) lies in Aries, rising around 10.30 pm.

• Set your alarm for a good view of **Mercury**. The innermost planet reaches greatest western elongation on 9 August, and is putting on the first of two excellent morning apparitions (the other is in November). Look very low in the east between 5 and 6 am, to spot this elusive world, which increases in brightness from +1.9 to −1.7 during August, though it will be easiest to spot mid-month when highest in the morning twilight.

• **Venus** and **Mars** are too close to the Sun to be visible in August.

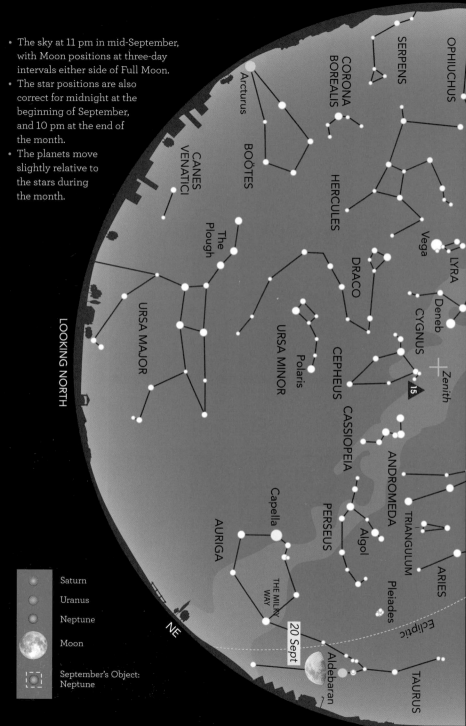

- The sky at 11 pm in mid-September, with Moon positions at three-day intervals either side of Full Moon.
- The star positions are also correct for midnight at the beginning of September, and 10 pm at the end of the month.
- The planets move slightly relative to the stars during the month.

WEST

OPHIUCHUS

SERPENS

CORONA BOREALIS

Arcturus

CANES VENATICI

BOÖTES

HERCULES

DRACO

Vega

LYRA

Deneb

CYGNUS

The Plough

Zenith

15

URSA MAJOR

URSA MINOR

Polaris

CEPHEUS

CASSIOPEIA

ANDROMEDA

TRIANGULUM

LOOKING NORTH

Algol

PERSEUS

ARIES

Capella

Pleiades

AURIGA

THE MILKY WAY

Ecliptic

NE

20 Sept

Aldebaran

TAURUS

Saturn

Uranus

Neptune

Moon

September's Object: Neptune

EAST

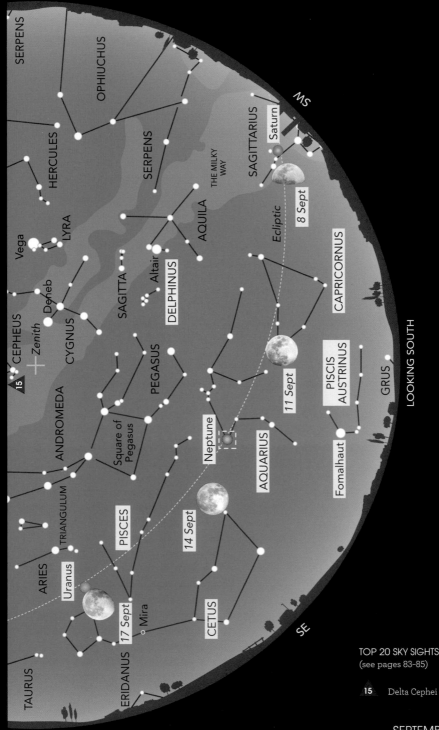

WEST

SERPENS

OPHIUCHUS

HERCULES

LYRA

Vega

CEPHEUS

Zenith

15

Deneb

CYGNUS

ANDROMEDA

TRIANGULUM

ARIES

TAURUS

ERIDANUS

EAST

SERPENS

SAGITTA

Altair

DELPHINUS

AQUILA

THE MILKY WAY

PEGASUS

Square of Pegasus

Neptune

PISCES

Uranus

17 Sept

Mira

CETUS

14 Sept

AQUARIUS

Ecliptic

8 Sept

SAGITTARIUS

Saturn

MS

CAPRICORNUS

11 Sept

PISCIS AUSTRINUS

Fomalhaut

GRUS

LOOKING SOUTH

SE

SEPTEMBER

TOP 20 SKY SIGHTS
(see pages 83–85)

15   Delta Cephei

It's not often we feature the most distant planet, but this month **Neptune** is not only at its closest but it also performs a rare pas-de-deux with a leading star in **Aquarius**. And the Water Carrier is just one of a host of dim, watery constellations in this month's celestial tableau: others include **Cetus** (the Sea Monster), **Capricornus** (the Sea Goat), **Pisces** (the Fishes), **Piscis Austrinus** (the Southern Fish) and **Delphinus** (the Dolphin).

SEPTEMBER'S CONSTELLATION

Piscis Austrinus (the Southern Fish) lies low in the south-west. It's hardly a compelling constellation. But it holds fond childhood memories for both of us, because this is the only time of year that we got to see its brightest star, **Fomalhaut**: the southernmost first-magnitude star visible from Britain.

Its name – derived from the Arabic – means 'the mouth of the whale', though one of its nicknames seems more appropriate: 'the lonely star of autumn'.

But Fomalhaut isn't really lonely. It has a disc of debris in orbit about it, with the potential to form a planetary family. And in 2008, the Hubble Space Telescope captured an image of one of its worlds – the first to be seen beyond the Solar System. It has been given the name Dagon (an ancient fertility god).

SEPTEMBER'S OBJECT

With Pluto being demoted to a mere 'ice dwarf', **Neptune** is officially the most remote planet in our Solar System. It lies 4,500 million kilometres out – 30 times the Earth's distance – in the twilight zone of our family of worlds, and takes nearly 165 years to circle the Sun.

Neptune is at its closest this year on **10 September**, and visible through a small telescope in Aquarius. But you need a spaceprobe to get up close and personal to the gas giant planet. In 1989, Voyager 2 discovered a turquoise world 17 times heavier than Earth, cloaked in clouds of methane and ammonia.

The most distant planet has a family of 14 moons, including Triton, which boasts erupting ice volcanoes. And the world is encircled by very faint rings of dusty debris.

For a world so far from the Sun, Neptune is amazingly frisky. Its core blazes at nearly 7000 °C: hotter than the Sun's surface. This internal heat drives dramatic storms, and winds of 2000 kilometres per hour – the fastest in the Solar System.

SEPTEMBER'S TOPIC: BIRTH OF THE MOON

The Earth–Moon system is unique in the Solar System. Compared to their planets, most moons are puny; our Moon is over a quarter the size of Earth. So where did it come from? How was it created?

By analysing rocks brought back by the Apollo astronauts, scientists have discovered an astonishing similarity between moonrocks and those on the surface of the Earth. They've come to a dramatic conclusion: that the Moon was literally blasted out of our planet.

Our companion world is the result of 'The Big Splash' – a cataclysm that befell the Earth over four billion years

Nik Szymanek captured this ghostly image of the Zodiacal Light from the Rocque de Los Muchachos Observatory, La Palma, on 3 June 2011. He used a Canon 5D MkII DSLR and Canon 28mm lens at f/2.8. Nik made 20-second exposures at ISO 1600. The dome in the foreground houses the UK's William Herschel Telescope.

ago. Then, the Solar System was a dangerous place; the young worlds had not settled down. A wayward body headed towards Earth, hell-bent on destroying our planet. The Mars-sized body hit its target, almost splitting apart our fledgling planet. But the larger Earth held its ground, pulling itself back together as a globe of molten lava.

The impact left a permanent legacy. Incandescent drops splashed into space, forming a super-hot ring around the Earth. The cooling droplets gradually came together, to create our Moon. When you look at the tranquil face of the Moon today, it's hard to believe that it had such a traumatic birth – in fire.

SEPTEMBER'S PICTURE

On a crystal-clear night – away from streetlights, or even the faintest moonlight – you may see a ghostly triangle of light at sunrise or sunset, extending upwards from the horizon. Lucky you! You've spotted the **Zodiacal Light**; a rare sight in today's light-polluted skies.

The Zodiacal Light is at its best in autumn (mornings), and spring (evenings), when the Zodiac – the line followed by the planets across the sky – rises at a steep angle from the horizon. It consists of a fog of tiny particles shed by ancient comets held in the thrall of mighty Jupiter, which scatter light from the Sun.

The 12th-century Persian astronomer and poet Omar Khayyam knew this 'false dawn' well. It was always an excuse for a glass of wine!

OBSERVING TIP

When you first go out to observe, you may be disappointed at how few stars you can see in the sky. But wait for 20 minutes, and you'll be amazed at how your night vision improves. One reason for this 'dark adaption' is that the pupil of your eye grows larger. More importantly, in dark conditions the retina of your eye builds up bigger reserves of rhodopsin, the chemical that responds to light.

SUNDAY	MONDAY	TUESDAY	WEDNESDAY	THURSDAY	FRIDAY	SATURDAY
1	2	3	4	5 Moon near Jupiter and Antares	6 4.10 am First Quarter Moon; Neptune very near phi Aquarii	7
8 Moon near Saturn	9	10 Neptune opposition	11	12	13	14 5.33 am Full Moon
15	16	17	18	19 Moon near Pleiades and Hyades	20 Moon near Pleiades and Hyades (am)	21
22 3.41 am Last Quarter Moon	23 Autumn Equinox	24 Moon near Castor and Pollux (am)	25 Moon near Praesepe (am)	26 Moon near Regulus (am)	27	28 7.26 pm New Moon
29	30					

SPECIAL EVENTS

- **5 September:** the bright 'star' to the left of the Moon is Jupiter (Chart 9a).
- **6 September:** Neptune lies almost in front of the star phi Aquarii; a memorable sight in a telescope (see Planet Watch).
- **8 September:** The Moon lies close to Saturn (Chart 9a).
- **10 September:** Neptune is opposite to the Sun in the sky and at its closest to Earth this year, 4328 million km away.
- **Night of 19/20 September:** the waning Moon lies below the Pleiades, and during the night moves towards the Hyades and Aldebaran.
- **23 September, 8.50 am:** nights become shorter than days as the Sun moves south of the Equator at the Autumn Equinox.
- **24 September:** in the morning sky, the crescent Moon lies beneath the twin stars Castor and Pollux (Chart 9b).
- **25 September:** the Moon lies to the left of the star cluster Praesepe before dawn (Chart 9b).
- **26 September:** look to the east before dawn to spot the thin crescent Moon above Regulus (Chart 9b).

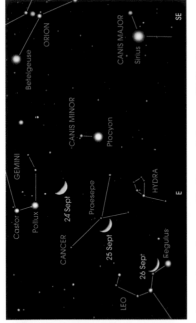

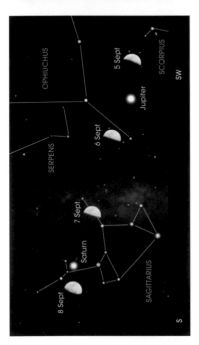

9a 5–8 September, 10 pm. The Moon with Jupiter and Saturn.

9b 24–26 September, 4 am. The Moon with Castor and Pollux; Praesepe; and Regulus

- **Jupiter** is brilliant in the evening sky, blazing at magnitude −2.1 low in the south-west in Ophiuchus. The giant planet sets about 10.30 pm.

- You'll find **Saturn** in the south, at magnitude +0.4 in Sagittarius and setting around 0.30 am.

- This month it's the outermost planet's time to shine! For starters, **Neptune** is at opposition on 10 September (see Special Events) so it's visible throughout the whole night, at its maximum brightness of magnitude +7.8. Though the planet is too dim to see with the naked eye, you can scoop it up in binoculars, and it's an easy target with a small telescope. The problem is knowing exactly where to point your instrument to discern Neptune among all the faint stars.

- This month, though, it's easy: Neptune is very close to a star bright enough to appear on our charts, phi Aquarii (magnitude +4.2). Between 3 and 9 September, the 'star' closest to phi Aquarii (and 25 times fainter) is Neptune. And on 6 September, Neptune is only half an arcminute to the right of phi Aquarii – less than the apparent size of Jupiter. You'll be hard pushed to separate the two in binoculars, and a telescope will reveal a stunning red and green 'double star'.

- **Uranus** rises at 8.30 pm, at magnitude +5.7 in Aries.

- **Mercury, Venus** and **Mars** are lost in the Sun's glare this month.

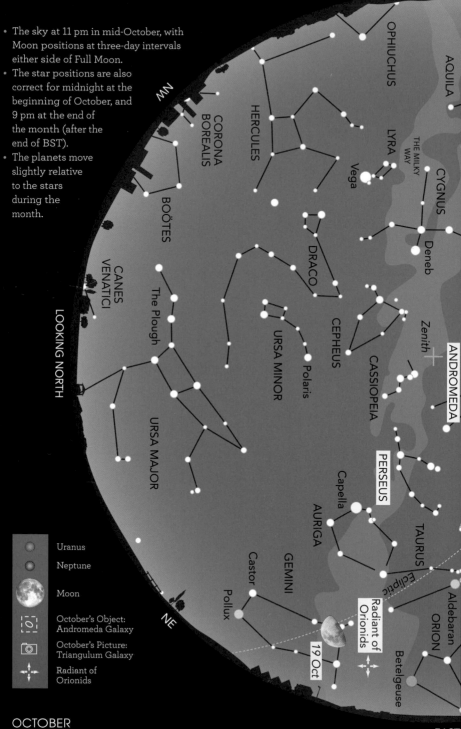

- The sky at 11 pm in mid-October, with Moon positions at three-day intervals either side of Full Moon.
- The star positions are also correct for midnight at the beginning of October, and 9 pm at the end of the month (after the end of BST).
- The planets move slightly relative to the stars during the month.

WEST

NW

LOOKING NORTH

NE

EAST

OPHIUCHUS

AQUILA

HERCULES

CORONA BOREALIS

LYRA

Vega

THE MILKY WAY

CYGNUS

Deneb

BOÖTES

DRACO

CEPHEUS

Zenith

ANDROMEDA

CANES VENATICI

The Plough

URSA MINOR

Polaris

CASSIOPEIA

PERSEUS

URSA MAJOR

Capella

AURIGA

TAURUS

Castor

GEMINI

Pollux

Radiant of Orionids

19 Oct

Ecliptic

Aldebaran

ORION

Betelgeuse

Uranus

Neptune

Moon

October's Object: Andromeda Galaxy

October's Picture: Triangulum Galaxy

Radiant of Orionids

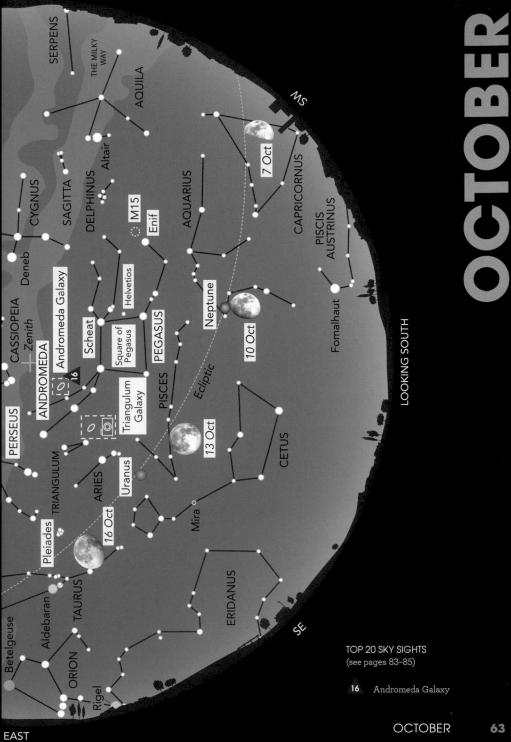

SERPENS

THE MILKY WAY

AQUILA

CYGNUS

SAGITTA

DELPHINUS

Altair

Deneb

CASSIOPEIA

Zenith

ANDROMEDA

Andromeda Galaxy

PERSEUS

TRIANGULUM

Pleiades

Aldebaran

TAURUS

Betelgeuse

ORION

Rigel

M15

Enif

Scheat

Helvetios

Square of Pegasus

PEGASUS

PISCES

Triangulum Galaxy

16

ARIES

Uranus

16 Oct

Mira

CETUS

ERIDANUS

7 Oct

MS

CAPRICORNUS

AQUARIUS

Neptune

10 Oct

PISCIS AUSTRINUS

Fomalhaut

Ecliptic

13 Oct

SE

TOP 20 SKY SIGHTS
(see pages 83–85)

16 Andromeda Galaxy

The glories of October's skies can best be described as 'subtle'. The barren **Square of Pegasus** dominates the southern sky, with **Andromeda** attached to his side. But the dull autumn constellations are already being faced down by the brilliant lights of winter, spearheaded by the beautiful star cluster of the **Pleiades**.

OCTOBER'S CONSTELLATION

Pegasus is little more than a large, empty square of four medium-bright stars. But our ancestors managed to see an upside-down winged horse here. In legend, Pegasus sprang from the blood of Medusa the Gorgon when **Perseus** severed her head.

A red giant star 200 times wider than the Sun, **Scheat** is nearing the end of its life and varies in brightness by a magnitude as it pulsates irregularly. **Enif** (the nose) is an orange supergiant, with a faint blue companion visible in a small telescope, or even good binoculars.

OBSERVING TIP

The Andromeda Galaxy is often described as the furthest object 'easily visible to the unaided eye'. (The Triangulum Galaxy, this month's Picture, is far from easy to spot!) It can be a bit elusive – especially if you are suffering from light pollution. The trick is to memorise Andromeda's pattern of stars, and then to look slightly to the *side* of where you expect the galaxy to be. This technique – called 'averted vision' – causes the image to fall on the outer region of your retina, which is more sensitive to light than the central region that's evolved to discern fine details. You'll certainly need averted vision to eyeball the Triangulum Galaxy. The technique is also crucial when you want to observe the faintest nebulae or galaxies through a telescope.

Next to Enif – and Pegasus's best-kept secret – is the beautiful globular cluster **M15**. You'll need a telescope for this one. M15 is 33,000 light years away, and contains about 100,000 densely-packed stars.

And Pegasus contains the first planet to be discovered beyond our Solar System, orbiting the star **Helvetios** (51 Pegasi), which is just visible to the unaided eye. The planet has been named Dimidium (meaning 'half' in Latin – it's half the mass of Jupiter).

OCTOBER'S OBJECT

Take the advantage of autumn's new-born darkness to pick out our neighbour in the Universe, the **Andromeda Galaxy** (catalogued as M31 in Charles Messier's list of fuzzy patches, see April's Topic).

Visible to the unaided eye, the Andromeda Galaxy covers an area four times bigger than the Full Moon. Like our Milky Way, it is a beautiful spiral shape, but – alas – it's presented to us almost edge-on. Even a small telescope won't reveal much detail.

The Andromeda Galaxy lies 2.5 million light years away: it's similar to the Milky Way, but larger. It also hosts two bright companion galaxies – just as our own Galaxy does – as well as a flotilla of orbiting dwarf galaxies.

Unlike other galaxies, which are receding from us in the expanding Universe, Andromeda is approaching the Milky Way. The two galaxies will merge in

From his home in Shropshire, Pete Williamson captured this image of M33 with the 430 mm f/4.5 Planewave iTelescope.net T21 in New Mexico, equipped with a SBIG 4008 x 2672 pixel CCD. Pete used an exposure of 4 x 300 seconds through R, V, B and H-alpha filters, and processed the image in Pixinsight 1.8.

about four billion years' time. The pile-up will create a giant elliptical galaxy – nick-named Milkomeda – devoid of gas and dominated by ancient red giant stars.

OCTOBER'S TOPIC: DARK MATTER

We think of the Universe as being luminous: alight with stars, nebulae and galaxies. But nothing could be further from the truth. In recent years, astronomers have discovered that almost 90 per cent of the matter in the Cosmos is invisible – taking the form of mysterious dark matter.

Suspicions were first aroused in 1933, when the great Swiss astronomer Fritz Zwicky found that galaxies in the Coma Cluster were moving unexpectedly fast. Something was exerting 400 times more gravity than all the galaxies in the cluster, and Zwicky called it 'dark matter'.

Later, researchers found that stars and gas in the outer regions of spiral galaxies were spinning round much faster than predicted. Again, dark matter in the galaxy must be holding them in.

Without knowing the nature of dark matter, it's hard to know how to find it – it's like looking for a black cat in a coal cellar! At the moment, the best bet is that dark matter is made up of WIMPs – a wonderful acronym for 'weakly interacting massive particles'. In theory, WIMPs were created in the Big Bang, and should still be around today.

Physicists are competing to winkle out these bizarre particles, including a group working deep in a potash mine in Yorkshire. Boulby Mine is over a kilometre below the ground, where the dark-matter detectors are shielded from radiation that could confuse the results.

OCTOBER'S PICTURE

The **Triangulum Galaxy** (M33) is the third-largest member of our Local Group. Much smaller and more unformed than its neighbour spirals – Andromeda and the Milky Way – it is nevertheless a veritable hotbed of star formation. Lying three million light years distant, M33 may be the furthest object visible to the unaided eye. Observers in desert locations claim to have seen this sprawled-out spiral of some 40 billion stars in pitch-black skies. Over to you: any reports from the UK?

SUNDAY	MONDAY	TUESDAY	WEDNESDAY	THURSDAY	FRIDAY	SATURDAY
		1	2	3 Moon near Jupiter	4	5 5.47 pm First Quarter Moon near Saturn
6	7	8	9	10	11	12
13 10.08 pm Full Moon	14	15	16	17 Moon in Hyades, near Aldebaran	18	19
20 Mercury E elongation; Moon near Castor and Pollux	21 1.39 pm Last Quarter Moon; Orionids	22 Orionids (am); Moon occults Praesepe (am)	23 Moon near Regulus (am)	24 Moon near Regulus (am)	25	26 Moon near Mars (am)
27 British Summer Time ends	28 3.38 am New Moon; Uranus opposition	29 Moon very near Venus	30 Moon between Venus and Jupiter	31 Moon near Jupiter		

⭐ SPECIAL EVENTS

- **3 October:** the crescent Moon lies just to the right of Jupiter, low in the south after sunset.
- **5 October:** Saturn lies right next to the First Quarter Moon, down in the south.
- **17 October:** as the Moon rises, about 8.30 pm, it's swimming through the Hyades just above Aldebaran.
- **Night of 21/22 October:** maximum of the Orionid **meteor shower**, when debris from Halley's Comet enters Earth's atmosphere. Observe in the late evening and overnight, looking away from the Moon rising in the east.
- **22 October, 4–7 am:** the Moon moves right in front of Praesepe, occulting many of its stars – a lovely sight in binoculars or a small telescope (Chart 10a).
- **27 October, 2 am:** the end of British Summer Time for this year, as clocks go back by an hour.
- **28 October:** Uranus is opposite to the Sun in the sky and at its closest to Earth this year, 2817 million km away.
- **29 October:** look very low in the dusk twilight, to spot the thinnest crescent Moon just above Venus; with binoculars, you may pick out Mercury below Venus (Chart 10b).
- **30 October:** a thin crescent Moon lies between Venus (right) and Jupiter (left) (Chart 10b).
- **31 October:** the crescent Moon lies very near Jupiter, with Venus to the lower right (Chart 10b).

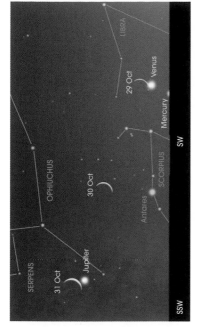

10a 22 October, 6 am. The Moon occults Praesepe.

10b 29-31 October, 5 pm. The crescent Moon with Venus, Mercury and Jupiter.

Uranus

- Right at the end of October, look low in the south-west after sunset to catch the planet **Venus** re-appearing in the evening sky: at magnitude –3.9, it's setting about 5.30 pm.

- Sweep the sky to the lower left of Venus with binoculars, and you may just catch **Mercury** tightly hugging the horizon and setting a few minutes earlier: after greatest eastern elongation on 20 October, Mercury fades from magnitude 0 to +0.6 at the end of the month.

- Lording it over the evening sky, **Jupiter** shines brilliantly in Ophiuchus at magnitude –2.0, setting around 8.30 pm.

- **Saturn**, in Sagittarius, sets about 10.30 pm. It's ten times fainter than Jupiter, at magnitude +0.5. It's followed by dim **Neptune** (magnitude +7.8) which lies in Aquarius sets around 4 am.

- **Uranus** is at opposition on 28 October (see Special Events) and is visible all night long in Aries: the planet reaches its maximum brightness of the year at magnitude +5.7, so it's just visible to the naked eye and can easily be seen with binoculars.

- In the dawn sky, look out for **Mars** in the east, rising around 6 am. The Red Planet lies in Virgo, at magnitude +1.8, and becomes more apparent during October as dawn comes ever later.

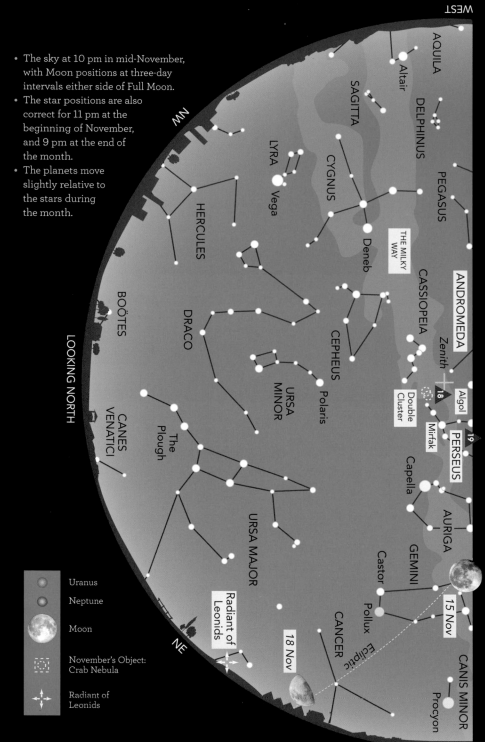

- The sky at 10 pm in mid-November, with Moon positions at three-day intervals either side of Full Moon.
- The star positions are also correct for 11 pm at the beginning of November, and 9 pm at the end of the month.
- The planets move slightly relative to the stars during the month.

WEST

AQUILA

Altair

SAGITTA

DELPHINUS

LYRA

Vega

CYGNUS

PEGASUS

THE MILKY WAY

NW

Deneb

CASSIOPEIA

HERCULES

ANDROMEDA

Zenith

CEPHEUS

Double Cluster

Algol

18

BOÖTES

DRACO

URSA MINOR

Polaris

Mirfak

PERSEUS

19

LOOKING NORTH

CANES VENATICI

The Plough

Capella

Uranus

Neptune

Moon

November's Object: Crab Nebula

Radiant of Leonids

URSA MAJOR

AURIGA

Castor

GEMINI

Pollux

15 Nov

Radiant of Leonids

18 Nov

CANCER

Ecliptic

CANIS MINOR

Procyon

NE

EAST

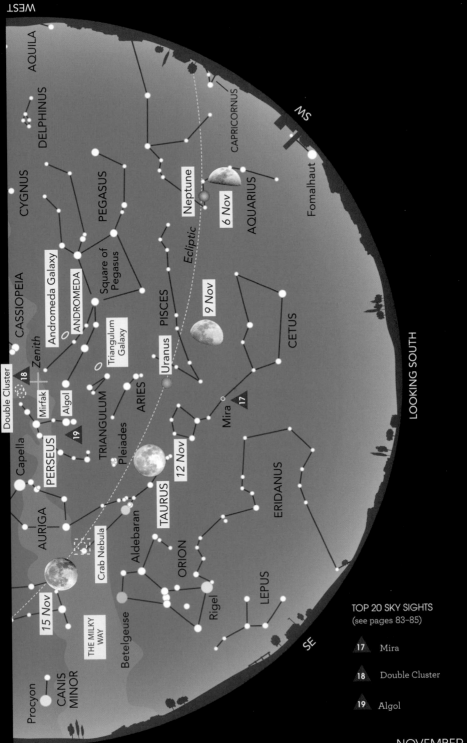

AQUILA

DELPHINUS

CYGNUS

PEGASUS

CASSIOPEIA

Andromeda Galaxy

ANDROMEDA

Square of
Pegasus

Zenith

Triangulum
Galaxy

Double Cluster

18

Mirfak

Algol

19

Capella

PERSEUS

TRIANGULUM

ARIES

Pleiades

AURIGA

Crab Nebula

Aldebaran

TAURUS

12 Nov

ORION

THE MILKY
WAY

15 Nov

Betelgeuse

Rigel

Procyon

CANIS
MINOR

EAST

SE

LEPUS

ERIDANUS

Mira

17

CETUS

PISCES

Uranus

Neptune

9 Nov

6 Nov

AQUARIUS

Ecliptic

CAPRICORNUS

MS

Fomalhaut

LOOKING SOUTH

TOP 20 SKY SIGHTS
(see pages 83–85)

17 Mira

18 Double Cluster

19 Algol

A daytime treat hits the headlines this month, as Mercury crosses the face of the Sun – see below for tips on observing this rare event safely. Low in the twilight glow, brilliant Venus gets up close and personal to second-brightest planet Jupiter. And, after dark, the **Milky Way** rears up and provides a stunning inside perspective on the huge Galaxy that is our home in space.

NOVEMBER'S CONSTELLATION

Perseus is one of the best-loved constellations in the northern night sky. It never sets from Britain, and is packed with celestial goodies. In legend, Perseus was the superhero who slew Medusa, the Gorgon. Its brightest star, **Mirfak** ('elbow' in Arabic), lies 510 light years away and is 5000 times more luminous than our Sun.

But the 'star' of Perseus has to be **Algol** (whose name stems from the Arabic al-Ghul – the 'Demon Star'). It represents the eye of Medusa – and it winks. Its variations were first reported by 18-year-old John Goodricke, a profoundly deaf amateur astronomer. He correctly surmised that a fainter star eclipses a brighter one (there are actually three stars in the system), so that Algol dims from magnitude +2.1 to +3.4 over a period of 2.9 days.

OBSERVING TIP

The event of the month is the transit of Mercury, but be very careful how you observe it. NEVER look at the Sun directly, with your naked eyes or – especially – with a telescope or binoculars: it could well blind you permanently. Project the Sun's image through a small telescope onto a piece of white card. Or – if you want the real biz – get some solar binoculars or a solar telescope, with filters that guarantee a safe view. Check the web for details.

Another pair of gems is the **Double Cluster**, h and chi Persei. Visible to the unaided eye, the duo is a sensational sight in binoculars. Some 7500 light years distant, the clusters are made of bright young blue stars. Both are a mere 12 million years old (compare this to our Sun, which has notched up 4.6 *billion* years so far!).

NOVEMBER'S OBJECT

Just above the 'lower horn' of **Taurus** (the Bull), Chinese astronomers witnessed the appearance of a brilliant 'guest star' in 1054. Visible in daylight for 23 days, the supernova remained in the night sky for nearly two years. But this was no stellar debutante – it was an old star on the way out, exploding because it was overweight.

Today, we see the remnants of this supernova as the **Crab Nebula** – named by 19th-century Irish astronomer, the Third Earl of Rosse, because it resembled a crab's pincers. The expanding debris now measures 11 light years across. You can just make out the Crab Nebula through a small telescope, but it is small and faint (magnitude +8.4).

At the centre of the Crab Nebula is the core of the dead star, which has collapsed to become a pulsar. This tiny, but super-dense object – only the size of a city, but with the mass of the Sun – is spinning around furiously at 30 times a second and emitting beams of radiation like a lighthouse.

NOVEMBER'S TOPIC: TRANSIT OF MERCURY

On 11 November, we're treated to the rare sight of our innermost planet crossing the face of the Sun. The transit of **Mercury** begins at 12.35 pm; it's halfway across at 3.20 pm, and the planet is still in transit at sunset (about 4.30pm).

The planet's silhouette looks like a small, sharp blob – unlike the fuzzy blur of a sunspot. When we say small, we mean very small. If you've seen a transit of Venus, you'll notice the difference! Venus is much closer to the Earth than Mercury, and larger. But at mid-transit this month, Mercury is bang in the centre of the Sun's disc, so you won't miss it (Chart 11a on page 73).

But before you observe the transit, we need to put in a warning: DO NOT LOOK AT THE SUN DIRECTLY; see the Observing Tip for details of safe viewing.

In the past, transits of the innermost planets – Mercury and Venus – were a way of gauging the size of our Solar System. By observing how quickly these worlds crossed the disc of our local star, astronomers could work out how far the Earth lies from the Sun. Pierre Gassendi first observed a transit of Mercury on 7 November 1631, while Captain Cook got to see a transit of Mercury on 9 November 1769 on his pioneering visit to New Zealand.

NOVEMBER'S PICTURE

The last time we saw a transit of Mercury was in 2016, when much of Britain was blessed with fine weather, enabling Dave Ratledge to capture this atmospheric view. The planet is at lower right – not the sunspot group in the middle. Let's hope for good luck this time!

If we miss out, though, don't despair. Mercury transits our local star 13 or 14 times every century (you can expect to see about half of these from any particular location). So, if we're unlucky, just hold out until 13 November 2032.

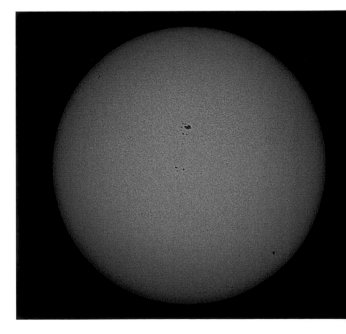

Dave Ratledge caught the transit of Mercury on 9 May 2016, using a Celestron C8 telescope with a f/6.3 focal reducer, a Thousand Oaks off-axis filter (aperture 75 mm) and Canon 60D camera. Dave took ten exposures at 1/250 second.

SUNDAY	MONDAY	TUESDAY	WEDNESDAY	THURSDAY	FRIDAY	SATURDAY
					1	2 Moon near Saturn
3	4 10.23 am First Quarter Moon	5	6	7	8	9
10 Mars near Spica (am)	11 Transit of Mercury	12 1.34 pm Full Moon	13 Moon occults Hyades, near Aldebaran	14 Moon occults Hyades, near Aldebaran (am)	15	16 Moon near Castor and Pollux
17 Leonid meteor shower	18 Leonid meteor shower (am)	19 9.11 pm Last Quarter Moon near Regulus	20	21	22	23
24 Moon near Mercury, Mars, Spica (am); Venus near Jupiter	25 Moon near Mercury (am)	26 3.06 pm New Moon	27	28 Mercury W elongation; Moon, Venus and Jupiter	29 Moon near Saturn	30

SPECIAL EVENTS

• **11 November:** this afternoon and evening, we are treated to the rare sight of Mercury moving across the face of the Sun (Chart 11a). The transit begins at 12.35 pm, and is still continuing when the Sun sets around 4.30 pm. DO NOT OBSERVE THE SUN DIRECTLY, BUT PROJECT ITS IMAGE. See this month's Topic and Observing Tip for more details.

• **Night of 13/14 November:** the Moon moves in front of the Hyades between moonrise and dawn, occulting many of its stars.

• **Night of 17/18 November:** maximum of the **Leonid meteor shower**. Occasionally over the past two centuries the Leonids have stormed the Earth with shooting stars, but we're not expecting any great fireworks this year as its parent body, Comet Tempel-Tuttle, is far away from us, near the orbit of Uranus.

• **24 November:** the two brightest planets, Venus and Jupiter, are close together in the south-west after sunset.

• **28 November:** as the sky grows dark (around 4.45 pm) look low in the south-west to see the lovely sight of the crescent Moon between Venus and Jupiter (Chart 11b).

• **29 November:** Saturn lies to the upper left of the crescent Moon (Chart 11b).

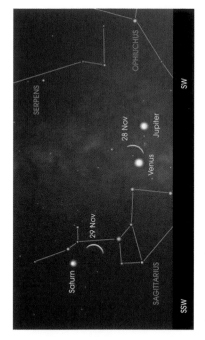

11a 11 November, 12.35–4.30 pm. Transit of Mercury, from ingress to sunset. Because of the Earth's rotation, Mercury's path appears to be curved. We've enlarged the planet for clarity – see this month's Picture for its true size.

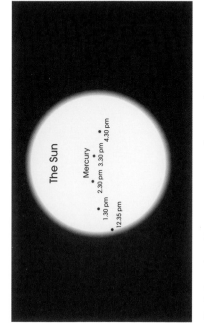

11b 28, 29 November, 4.45 pm. The crescent Moon with Venus, Jupiter and Saturn.

- It's an exciting month for planet-lovers, as the two brightest worlds come together in the evening sky. Brilliant **Venus** is steaming upwards from the south-western horizon after sunset, setting around 5 pm. At magnitude –3.9, the Morning Star is the brightest object in the night sky after the Moon, and it becomes more prominent as the evenings darken.

- At the start of November, you'll find second-brightest planet **Jupiter** well to the upper left of Venus, shining at magnitude –2.9 in Ophiuchus and setting about 6.30 pm. Venus – six times brighter than Jupiter – passes just a degree-and-a-half below the giant planet on 24 November. **Saturn** lies to the upper left of these canoodling planets, shining at magnitude +0.6

in Sagittarius, and setting around 7.30 pm.
- Outermost planet **Neptune** (magnitude +7.9) is in Aquarius, setting about 1 am. Following along is **Uranus**, lying in Aries: it shines at magnitude +5.7 and sets around 5.30 am.
- In the morning sky, **Mars** is rising about 5 am: at magnitude +1.7, the Red Planet lies in Virgo, and on 10 November passes

the constellation's brightest star, Spica.
- After its transit, **Mercury** joins Mars before dawn. The innermost planet appears above the south-eastern horizon around 18 November, brightening from magnitude +1.2 to –0.5 as it moves towards Mars and greatest western elongation on 28 November. By the end of the month Mercury is rising as early as 5.45 am.

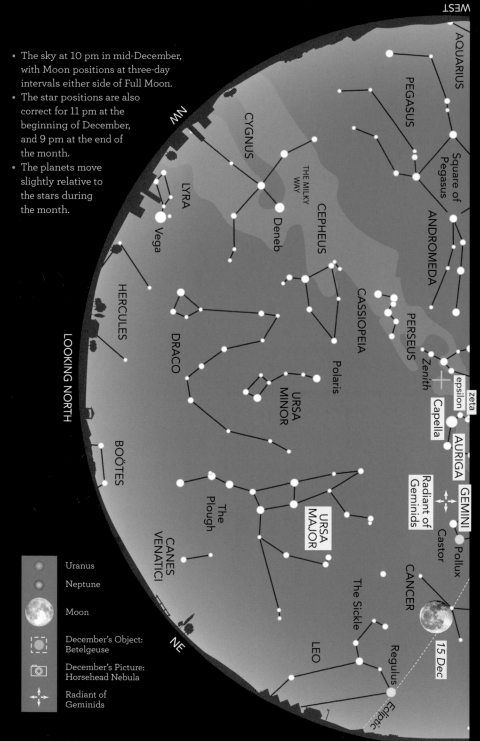

- The sky at 10 pm in mid-December, with Moon positions at three-day intervals either side of Full Moon.
- The star positions are also correct for 11 pm at the beginning of December, and 9 pm at the end of the month.
- The planets move slightly relative to the stars during the month.

WEST

AQUARIUS

PEGASUS

Square of Pegasus

ANDROMEDA

NW

CYGNUS

THE MILKY WAY

CEPHEUS

PERSEUS

LYRA

Vega

Deneb

CASSIOPEIA

Zenith

epsilon

zeta

Capella

AURIGA

HERCULES

DRACO

Polaris

URSA MINOR

Radiant of Geminids

GEMINI

Castor

Pollux

BOÖTES

The Plough

URSA MAJOR

CANCER

CANES VENATICI

The Sickle

LEO

Regulus

15 Dec

NE

Ecliptic

LOOKING NORTH

Uranus

Neptune

Moon

December's Object: Betelgeuse

December's Picture: Horsehead Nebula

Radiant of Geminids

EAST

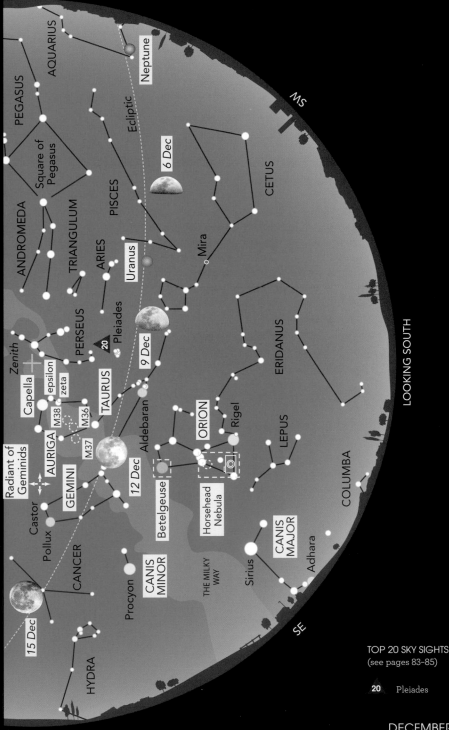

DECEMBER

AQUARIUS

PEGASUS

Neptune

Ecliptic

MS

Square of Pegasus

6 Dec

PISCES

CETUS

ANDROMEDA

TRIANGULUM

ARIES

Uranus

Mira

PERSEUS

Pleiades

20

9 Dec

ERIDANUS

Zenith

LOOKING SOUTH

Capella

epsilon

zeta

TAURUS

M38

M36

Aldebaran

ORION

Radiant of Geminids

AURIGA

M37

GEMINI

12 Dec

Rigel

LEPUS

Castor

Betelgeuse

Pollux

CANCER

Horsehead Nebula

COLUMBA

Procyon

CANIS MINOR

THE MILKY WAY

CANIS MAJOR

Adhara

Sirius

SE

15 Dec

TOP 20 SKY SIGHTS
(see pages 83–85)

HYDRA

20 Pleiades

We have a Christmas 'star': glorious Venus, swinging round the Sun into the evening sky. We're also treated to the shooting stars from the **Geminid** meteor shower, and the brilliant constellations of winter: **Orion**, his hunting dogs **Canis Major** and **Canis Minor**, **Taurus** (the Bull), the hero twins of **Gemini**, and the charioteer **Auriga** almost overhead.

DECEMBER'S CONSTELLATION

Auriga (the Charioteer) is named after the lame Greek hero Erichthoneus, who invented the four-horse chariot. The constellation's roots date way back to the ancient Babylonians, who saw Auriga as a shepherd's crook.

Capella, the sixth-brightest star in the sky, means 'the little she-goat'; but there's nothing modest about this giant yellow star. In fact, it consists of two substantial stars, each ten times wider than the Sun and about 75 times brighter, orbiting more closely than the Earth circles the Sun. More controversially, this pair may also be orbited by two faint red dwarf stars.

Nearby, you'll find a tiny triangle of stars nicknamed 'the Kids' (Haedi). Two are eclipsing binaries: stars that change in brightness because a companion passes in front. **Zeta** Aurigae is an orange star eclipsed every 972 days by a blue partner.

Epsilon Aurigae is a weirdo. Every 27 years, it suffers two-year-long eclipses, caused by a dark disc of material as big as the orbit of Jupiter. No two eclipses are the same, and there are tantalising hints of giant proto-planets within the disc.

Also, bring out those binoculars (better still, a small telescope) – to sweep within the 'body' of the Charioteer to find three very pretty open star clusters, **M36**, **M37** and **M38**.

DECEMBER'S OBJECT

Known to generations of schoolkids as 'Beetlejuice', **Betelgeuse** is one of the biggest and most luminous stars known. If placed in the Solar System, it would swamp the planets all the way out to the asteroid belt – possibly as far as Jupiter. It's over 100,000 times brighter than the Sun.

Almost 1000 times wider than the Sun, Betelgeuse is a serious red giant – a star close to the end of its life. It's one of just a few stars to be imaged as a visible disc from Earth, by specialised telescopes. Betelgeuse has suffered middle-age spread as frenetic nuclear reactions in the star's core have forced its outer layers to swell and cool. The star also fluctuates slightly in brightness as it tries to get a grip on its billowing gases.

The origin of the star's name is a

OBSERVING TIP

Venus is a real treat this month. Through a small telescope, you'll be able to make out its gibbous shape. But don't wait for the sky to get totally dark. Seen against a black sky, the cloud-wreathed world is so brilliant it's difficult to make out any details. You're best off viewing Venus soon after the Sun has set, when the Evening Star first becomes visible in the twilight glow. Through a telescope, the planet then appears less dazzling against a pale blue sky.

mystery. It comes from Arabic, and literally could mean 'the armpit of the sacred one'! But scholars now think the initial 'B' should really be a 'Y', and Betelgeuse is – more boringly – 'the giant's hand'.

Whatever its name means, Betelgeuse will exit the cosmic scene in a spectacular supernova explosion. As a result of the breakdown of nuclear reactions at its heart, the star will explode – to shine as brightly in our skies as the Full Moon.

DECEMBER'S TOPIC: VENUS – TWIN PLANET GONE WRONG

Venus – the planet of love – is resplendent in our evening skies this month. So brilliant and beautiful, she can even cast a shadow in a really dark, transparent sky. Her purity and lantern-like luminosity are beguiling – but looks are deceptive.

Earth's twin in size, Venus could hardly be more different from our warm, wet world. The reason for the planet's brilliance is the highly reflective clouds that cloak its surface: probe under these palls of sulphuric acid hanging in an atmosphere of carbon dioxide, and you find a planet out of hell.

From his home in West Horndon, Essex, Nik Szymanek captured this image of the iconic Horsehead Nebula with a GSO 10 inch (250 mm) Ritchey-Chrétien telescope, Astro-Physics 0.67x reducer, QSI 583 WSG CCD camera and an Astrodon hydrogen-alpha filter, combining nine 600-second exposures. Nik incorporated other colours from an earlier image taken with a Pentax 74 mm refractor and Astronomik filters.

Volcanoes are to blame. They have created a runaway greenhouse effect that has made Venus the hottest and most poisonous planet in the Solar System. At 460°C, this world is hotter than an oven. The pressure at its surface is around 90 Earth-atmospheres. So – if you visited Venus – you'd be simultaneously roasted, crushed, corroded and suffocated!

DECEMBER'S PICTURE

The celestial chess-piece of the Universe: Orion's **Horsehead Nebula**. This dark cloud of dust and gas, seven light years from nose to mane, is poised to create a new generation of stars. Newly born stars behind it are already lighting up swirling fronds of gas which act as a backdrop to the distinctive sihouette.

SUNDAY	MONDAY	TUESDAY	WEDNESDAY	THURSDAY	FRIDAY	SATURDAY
1	2	3	4 6.58 am First Quarter Moon	5	6	7
8	9	10 Venus near Saturn; Moon near Pleiades and Hyades	11 Moon near Hyades and Aldebaren (am); Venus near Saturn	12 5.12 am Full Moon; Mars near Zubenelgenubi (am)	13 Geminids	14 Geminids; Moon near Castor and Pollux
15 Geminids (am)	16 Moon near Regulus	17	18	19 4.57 am Last Quarter Moon	20 Moon near Spica	21
22 Winter Solstice	23 Moon near Mars (am)	24 Moon near Antares (am)	25	26 5.13 am New Moon; solar eclipse	27	28 Moon near Venus
29 Moon near Venus	30	31				

SPECIAL EVENTS

- **10, 11 December:** Brilliant Venus passes Saturn low in the south-west after sunset (Chart 12a).
- **Nights of 13/14 and 14/15 December:** look out for the bright slowly moving shooting stars of the **Geminid meteor shower,** which – unusually – are debris not from a comet, but from an asteroid, called Phaethon. It's one of the most prolific annual showers, but the display this year is spoilt by bright moonlight.
- **22 December, 4.19 am:** the Winter Solstice, when the Sun reaches its lowest point in the heavens as seen from the northern hemisphere, giving us the shortest day and the longest night.
- **26 December:** an annular solar eclipse – where the Moon moves right in front of the Sun but leaves a ring of the brilliant solar surface visible – will be seen from a narrow track stretching from the Arabian peninsula, through southern India and Sri Lanka to Indonesia and Singapore. Much of Asia and Australia experiences a partial eclipse – but none of the eclipse is visible from the UK.
- **28, 29 December:** there's a beautiful tableau in the dusk sky, as the crescent Moon joins brilliant planet Venus (Chart 12b).

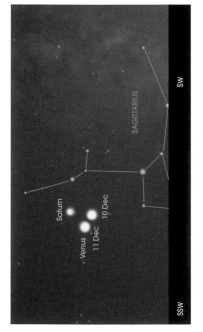

12a 10, 11 December, 4.30 pm. Conjunction of Venus and Saturn.

12b 28, 29 December, 4.30 pm. The crescent Moon with Venus.

There's a fascinating tango of planets in Sagittarius. As darkness falls, you can't miss brilliant **Venus** in the western sky, at a dazzling magnitude –4.0, and setting around 6 pm.

At the start of the month, **Jupiter** lies to its lower right. Seven times fainter, at magnitude –1.8, the giant planet sets about 5 pm. Meanwhile, **Saturn** is placed to the upper left of the Evening Star, setting around 6 pm: at magnitude +0.6, the ringworld is 40 times fainter than Venus.

As the days pass, Jupiter sinks rapidly into the twilight and disappears by the middle of the month. Meanwhile, Venus moves rapidly towards Saturn, and passes under the more distant planet on 10–11 December (Chart 12a).

Saturn drops into the dusk glow to depart the scene by the end of the year, while Venus rides ever higher.

- **Neptune** (magnitude +7.9) lies in Aquarius and sets around 11 pm, **Uranus**, in Aries at magnitude +5.7, is setting about 3.30 am.

- **Mars** rises around 5 am, shining at magnitude +1.6 in Libra. On the morning of 12 December, the Red Planet passes only 15 arcminutes from the double star Zubenelgenubi: a lovely sight in binoculars or a small telescope.

- At the beginning of the month, **Mercury** lies to the lower left of Mars, rather brighter at magnitude –0.6, and rising about 6 am. But it's dropping down in the sky, and disappears by mid-December.

Can you see the planets? We're amazed when people ask us that question: our closest cosmic neighbours are the brightest objects in the night sky after the Moon. Being so close, you can watch them getting up to their antics from night to night. Plus, there's planetary debris – the comets and meteors that are leftovers from the birth of the Solar System. All the info you need for observing them is here too.

THE INFERIOR PLANETS

A planet with an orbit that lies closer to the Sun than the orbit of Earth is known as *inferior*. Mercury and Venus are the inferior planets. They show a full range of phases (like the Moon) from the thinnest crescents to full, depending on their position in relation to the Earth and the Sun. The diagram below shows the various positions of the inferior planets. They are invisible when at *conjunction*, when they are either behind the Sun, or between the Earth and the Sun, and lost in the latter's glare.

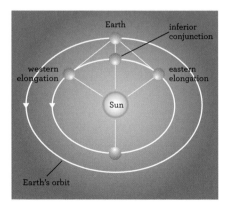

At eastern or western elongation, an inferior planet is at its maximum angular distance from the Sun. Conjunction occurs at two stages in the planet's orbit. Under certain circumstances, an inferior planet can transit across the Sun's disc at inferior conjunction.

Mercury

In the evening sky, Mercury is well visible in February–March, though you'll need to wait till mid-June for the best view of this elusive planet; its evening appearance in October is lost in the bright twilight sky. The innermost planet is difficult to spot in the dawn twilight at its April apparition, but it puts on good morning shows in August and November–December.

Maximum elongations of Mercury in 2019	
Date	Separation
27 February	18° east
11 April	28° west
23 June	25° east
9 August	19° west
20 October	25° east
28 November	20° west

Venus

After a brilliant appearance in the morning sky in January, Venus hugs the Sun for most of 2019, appearing low in the dawn twilight until July. It's an Evening Star from October through to the end of the year.

Maximum elongation of Venus in 2019	
Date	Separation
6 January	47° west

THE SUPERIOR PLANETS

The superior planets are those with orbits lying beyond that of the Earth: Mars, Jupiter, Saturn, Uranus and Neptune. The best time to observe a superior planet is when the Earth lies between it and the Sun. At this point it is at *opposition*.

Mars

For the first half of the year, the Red Planet is hanging around in the evening sky, gradually fading as the Earth pulls away. After passing behind the Sun on **2 September**, you'll catch Mars reappearing in the morning sky from October onwards.

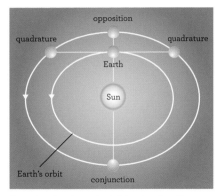

Superior planets are invisible at conjunction. At quadrature the planet is at right angles to the Sun as viewed from Earth. Opposition is the best time to observe a superior planet.

● Where to find Mars	
January to mid-February	Pisces
Mid-February to mid-March	Aries
Mid-March to mid-May	Taurus
Mid-May to mid-July	Gemini
October to November	Virgo
December	Libra

Jupiter

The giant planet kicks off 2019 in the morning sky; by May it's rising before midnight. Jupiter reaches opposition on **10 June**, and is prominent in the evening sky until mid-December. Throughout the year, it lies in Ophiuchus.

Saturn

The ringed planet is skulking low in the sky, in Sagittarius. Saturn is visible in the morning sky at the start of 2019 and can be seen all night when it reaches opposition on **9 July**. Saturn slips into the evening twilight at the very end of the year.

Uranus

Just perceptible to the naked eye, Uranus lies in Aries all year. It's visible in the evening sky from January to March, and re-emerges in the dawn sky in June. Uranus is at opposition on **28 October**.

Neptune

The most distant planet lies in Aquarius all year, and is at opposition on **10 September**. Neptune can be seen (though only through binoculars or a telescope) in January and February, and then from May until the end of the year.

SOLAR ECLIPSES

On **6 January**, people in the north-west Pacific, including Japan and parts of China, will witness a partial eclipse of the Sun; it's not visible from the UK.

A total eclipse of the Sun on **2 July** is visible from a narrow track that sweeps across the southern Pacific Ocean, Chile and Argentina. From the UK, we won't see any trace of the event.

On **26 December**, an annular solar eclipse will be seen from a narrow track through the Arabian peninsula, southern India and Sri Lanka to Indonesia and Singapore; not visible from the UK.

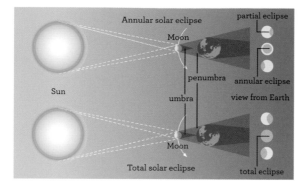

Where the dark central part (the umbra) of the Moon's shadow reaches the Earth, we see a total eclipse. People located within the penumbra see a partial eclipse. If the umbral shadow does not reach the Earth, we see an annular eclipse. This type of eclipse occurs when the Moon is at a distant point in its orbit and is not quite large enough to cover the whole of the Sun's disc.

LUNAR ECLIPSES

The total eclipse of the Moon on the night of **20/21 January** is the most stunning eclipse we'll see from the UK this year: an unusually large and bright supermoon is engulfed by the Earth's shadow.

On **16 July**, we have a partial eclipse of the Moon, visible from the UK: two-thirds of the Moon is obscured.

METEOR SHOWERS

Shooting stars, or *meteors,* are tiny specks of interplanetary dust, burning up in the Earth's atmosphere. At certain times of year, Earth passes through a stream of debris (usually left by a comet) and we see a *meteor shower.* The meteors appear to emanate from a point in the sky known as the *radiant.* Most showers are known by the constellation in which the radiant lies.

When watching meteors for a co-ordinated meteor programme, observers generally note the time, seeing conditions, cloud cover, their own location, the time and brightness of each meteor, and whether it was from the main meteor stream. It is also worth noting details of persistent afterglows (trains) and fireballs, and making counts of how many meteors appear in a given period.

COMETS

Comets are dirty snowballs from the outer Solar System. If they fall towards the Sun, its heat evaporates their ices to produce a gaseous head (*coma*) and sometimes dramatic tails. Although some comets are visible to the naked eye, you'll need a telescope to see the fine detail in the coma.

Hundreds of comets move round the Sun in small orbits. But many more don't return for thousands or even millions of years. Most comets are now discovered in professional surveys of the sky, but a few are still found by dedicated amateur astronomers.

In early January, Comet Wirtanen should be visible to the naked eye high in the northern sky, but later in the month you'll need binoculars to spot it. And watch out in case a brilliant new comet puts in a surprise appearance!

Dates of maximum for selected meteor showers	
Meteor Shower	Date of maximum
Quadrantids	3/4 January
Lyrids	22/23 April
Eta Aquarids	5/6 May
Perseids	12/13 August
Orionids	21/22 October
Leonids	17/18 November
Geminids	13/14 and 14/15 December

A new feature for 2019! We've always had our favourite sights in the night sky: and here they are in a season-by-season summary. It doesn't matter if you're a complete beginner, finding your way around the heavens with the unaided eye 👁 or binoculars 🔭; or if you're a seasoned stargazer, with a moderate telescope 📡. There's something here for everyone.

Each sky sight comes with a brief description, and a guide as to how you can best see it. Many of the most delectable objects are faint, so avoid moonlight when you go out spotting. Most of all, enjoy!

SPRING

Praesepe 👁 🔭 📡

Constellation: Cancer
Star Chart/Key: March; **5**
Type/Distance: Star cluster; 600 light years
Magnitude: +3.7
A fuzzy patch to the unaided eye; a telescope reveals many of its 1000 stars.

M81 and M82 🔭 📡

Constellation: Ursa Major
Star Chart/Key: March; **6**
Type/Distance: Galaxies; 12 million light years
Magnitude: +6.9 (M81); +8.4 (M82)
A pair of interacting galaxies: the spiral M81 appears as an oval blur, and the starburst M82 as a streak of light.

The Plough 👁

Constellation: Ursa Major
Star Chart/Key: April; **7**
Type/Distance: Asterism; 82–123 light years
Magnitude: Stars are roughly magnitude +2

The Plough

The seven brightest stars of the Great Bear form a large saucepan shape, called the Plough.

Mizar and Alcor 👁 🔭 📡

Constellation: Ursa Major
Star Chart/Key: April; **8**
Type/Distance: Double star; 80 light years
Magnitude: +2.3 (Mizar); +4.0 (Alcor)
The sky's classic double star, easily separated by the unaided eye: a telescope reveals Mizar itself is a close double.

Virgo Cluster 🔭 (difficult) 📡

Constellation: Virgo
Star Chart/Key: May; **9**
Type/Distance: Galaxy cluster; 54 million light years
Magnitude: Galaxies range from magnitude +9.4 downwards
Huge cluster of 2000 galaxies, best seen through moderate to large telescopes.

SUMMER

Antares 👁 🔭 📡

Constellation: Scorpius
Star Chart/Key: June; **10**
Type/Distance: Double star; 550 light years
Magnitude: +0.96
Bright red star close to the horizon. You can spot a faint green companion with a telescope.

M13 👁 🔭 📡

Constellation: Hercules
Star Chart/Key: June; **11**
Type/Distance: Star cluster; 22,200 light years
Magnitude: +5.8

M13

A faint blur to the naked eye, this ancient globular cluster is a delight seen through binoculars or a telescope. It boasts nearly a million stars.

Lagoon and Trifid Nebulae

Constellation: Sagittarius
Star Chart/Key: July; **12**
Type/Distance: Nebulae; 4000 light years (Lagoon); 5200 (Trifid)
Magnitude: +6.0 (Lagoon); +7.0 (Trifid)
While the Lagoon Nebula is just visible to the unaided eye, you'll need binoculars or a telescope to spot the Trifid. The two are in the same telescope field of view, and present a stunning photo opportunity.

Albireo

Constellation: Cygnus
Star Chart/Key: August; **13**
Type/Distance: Double star; 400 light years
Magnitude: Albireo A: +3.2; Albireo B: +5.1
Good binoculars reveal Albireo as being double. But you'll need a small telescope to appreciate its full glory. The brighter star appears golden; its companion shines piercing sapphire. It is the most beautiful double star in the sky.

Dumbbell Nebula

Constellation: Vulpecula
Star Chart/Key: August; **14**
Type/Distance: Planetary nebula; 1300 light years
Magnitude: +7.5
Visible through binoculars, and a lovely sight through a small/medium telescope, this is a dying star that has puffed off its atmosphere into space.

AUTUMN

Delta Cephei

Constellation: Cepheus
Star Chart/Key: September; **15**
Type/Distance: Variable star; 890 light years
Magnitude: +3.5 to +4.4, varying over 5 days 9 hours
The classic variable star, Delta Cephei is chief of the Cepheids – stars that allow us to measure distances in the Universe (their variability time is coupled to their intrinsic luminosity). Visible to the unaided eye, but you'll need binoculars for serious observations.

Andromeda Galaxy

Constellation: Andromeda
Star Chart/Key: October; **16**
Type/Distance: Galaxy; 2.5 million light years
Magnitude: +3.4
The biggest major galaxy to our own, the Andromeda Galaxy is easily visible to the unaided eye in unpolluted skies. Four times the width of the Full Moon, it's a great telescopic object and photographic target.

Andromeda Galaxy

Mira

Constellation: Cetus
Star Chart/Key: November; **17**
Type/Distance: Variable star; 300 light years
Magnitude: +3.5 to +10.1 over 332 days, although maxima and minima may vary
Mira is not nicknamed 'the Wonderful' for nothing. This distended red giant star is

alarmingly variable as it swells and shrinks
At its brightest, it's a naked-eye object;
binoculars may catch it at minimum; but
you need a telescope to monitor this star. Its
behaviour is highly unpredictable, and it's
important to keep logging it.

Double Cluster

Constellation: Perseus
Star Chart/Key: November; **18**
Type/Distance: Star clusters; 7500 light years
Magnitude: +3.7 and +3.8
A lovely sight to the unaided eye, these
stunning young star clusters are sensational
through binoculars or a small telescope.
They're a great photographic target.

Algol

Constellation: Perseus
Star Chart/Key: November; **19**
Type/Distance: Variable star; 90 light years
Magnitude: +2.1 – +3.4 over 2 days 21 hours
Like Mira, Algol is a variable star, but not an
intrinsic one. It's an 'eclipsing binary' – its
brightness falls when a fainter companion
star periodically passes in front of the
main star. It's easily monitored by the eye,
binoculars or a telescope.

WINTER

Pleiades

Constellation: Taurus
Star Chart/Key: December; **20**
Type/Distance: Star cluster; 400 light years
Magnitude: Stars range from magnitude +2.9
downwards
To the naked eye, most people can see six
stars in the cluster, but it can rise to 14 for the
keen-sighted. In binoculars or a telescope,
they are a must-see. Astronomers have
observed 1000 stars in the Pleiades.

Orion Nebula

Constellation: Orion
Star Chart/Key: January'
Type/Distance: Nebula; 1300 light years
Magnitude: +4.0
A striking sight even to the unaided eye, the

Orion Nebula

Orion Nebula – a star-forming region 24 light
years across – hangs just below Orion's belt.
Through binoculars or a small telescope, it is
staggering. A photographic must!

Betelgeuse

Constellation: Orion
Star Chart/Key: January; **2**
Type/Distance: Variable star; 640 light years
Magnitude: 0.0 – +1.3
Even with the unaided eye, you can see that
Betelgeuse is slightly variable over months,
as the red giant star billows in and out.

M35

Constellation: Gemini
Star Chart/Key: February; **3**
Type/Distance: Star cluster; 2800 light years
Magnitude: +5.3
Just visible to the unaided eye, this cluster of
around 2000 stars is a lovely sight through a
small telescope.

Sirius

Constellation: Canis Major
Star Chart/Key: February; **4**
Type/Distance: Double star; 8.6 light years
Magnitude: –1.47
You can't miss the Dog Star. It's the brightest
star in the sky! But you'll need a 150 mm
reflecting telescope (preferably bigger) to
pick out its +8.44 magnitude companion – a
white dwarf nicknamed 'the Pup'.

Sooner or later, every budding astronomer starts thinking about buying a telescope. These days the choice can be bewildering as there are hundreds of different models on the market. And, as with choosing a car, there is no single ideal telescope that suits everyone. So in this article I'll try to help narrow down the choice for you.

There are three main types of telescope: *refractors*, *reflectors* and *catadioptrics*. Each has its own strengths and weaknesses, and each tends to be more suited to a particular type of observing and indeed environment. To explain everything would take a book rather than an article, but here's a very quick guide to each type.

Refractors are the traditional type of telescope, with a lens at the top end, and you look through the bottom end. They are generally robust and need little maintenance, and are particularly useful in light-polluted areas where you need maximum contrast in the image. You can get short, stubby ones that are good for travel, but the cheaper versions tend to give images that have a bit of false colour around objects.

Reflectors use a mirror at the bottom of the tube rather than a lens at the top, so the eyepiece is at the upper end and you look sideways into the tube because there is a secondary mirror centrally placed to reflect the light out. Because the mirrors can tarnish and can get out of alignment, reflectors are less robust than refractors, but they are free from false colour and can be made in larger sizes. Unwanted light shining into the open-ended tube can be a problem in light-polluted areas. But you get the maximum telescope for your money with a reflector, usually of Newtonian optical design.

With both refractors and reflectors, the best image quality tends to be provided by the versions with longer tubes for their diameter. The crucial figure that all the ads refer to is the *focal ratio*, or *f-number*. The larger this figure, the easier it is to get high magnifications, but the more unwieldy the instrument. So an f/5 instrument will have a comparatively short tube and be more portable than an f/8 instrument, but will have optical shortcomings. These 'short-focus' instruments are more suited to low magnifications (up to about 100 times, say). But the lower the magnification the wider the field of view, so short-focus telescopes are better for looking the larger deep-sky objects such

The Sky-Watcher Evostar-90 on EQ2 mount is a budget 90 mm f/10 refractor on a basic non-driven equatorial mount for under £200. The same instrument is available on an altazimuth mount at lower cost, or you can add a motor drive to this version or obtain the same instrument on a heavier mounting.

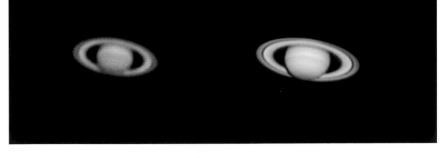

as nebulae, star clusters and nearby galaxies than planets or small deep-sky objects such as planetary nebulae.

High-quality refractors with what are known as ED or sometimes fluoride lenses, which use special types of glass, do combine good image quality with short tubes, but are considerably more expensive than normal refractors of the same size.

Catadioptrics, which are mostly either Schmidt-Cassegrain (SCT) or Maksutov optical designs, use a mirror at the bottom end and a corrector plate at the top end together with a secondary mirror in the tube to give a compact design that still gives high magnifications – at a price. They are quite robust and have closed tubes, and are widely used for planetary observing on account of their high power for their size. Though they have smaller fields of view than short-focus telescopes, they can still show you most deep-sky objects in one field of view.

WHAT SIZE?

As well as focal ratio, the other crucial figure is the aperture of the telescope – the diameter of the mirror or lens. These

Saturn photographed using a 130 mm Newtonian reflector (left) and 200 mm SCT (right). The larger the telescope, in general the brighter and more detailed the image – though turbulence in our atmosphere (known as 'seeing') often limits the amount of detail visible.

start at about 70 mm, and anything up to 130 mm is regarded as small, though by no means useless. You can see a lot with such scopes, given the right skies!

Telescopes with apertures from 130 mm to 200 mm are medium sized, and are used by a large number of amateur astronomers as their main instruments. They will show plenty of detail on planets, and have thousands of deep-sky objects within their grasp. Anything over 200 mm is getting quite large and serious, in terms of cost, performance and bulk. I wouldn't recommend such a large telescope for a beginner, though.

Celestron offer this 130 mm f/5 Newtonian reflector, the Astro-Fi 130, on a driven Go To altazimuth mount for under £500. It can be controlled via Wi-Fi from an Apple or Android app to find over 120,000 objects. A similar instrument is available on a basic manual equatorial mount for less than £200.

CHOOSING A MOUNTING

When choosing the right telescope for you, you also need to consider the type of mounting you get. The job of the mounting is to keep the telescope steady and to allow you to follow objects smoothly as the sky turns, which is important when you are magnifying the view maybe 100 times. Simple pan-and-tilt mechanisms as used on photographic tripods, for example, are far too clumsy.

Again there are several types, divided into *altazimuth* (which give right–left or up–down movements) and *equatorial*, which need to be aligned parallel to the Earth's axis so as to follow the sky. The most basic altazimuth mounts are used on the smallest and cheapest refractors and reflectors, and usually have limited ability to provide smooth movements, though some are better than others. Many budget refractors and reflectors are sold on simple manually driven equatorial mounts that require some getting used to and probably cause more problems for beginners than anything else. The advantage of an equatorial mount is that it allows you to follow objects through the sky using one tracking motion only, but this ability is often wasted if you don't know the sky well enough to set it up properly.

Small motorised altazimuth mounts are becoming increasingly common. These also need some setting up in advance, but once you've done this during an observing session they will keep objects in the field of view without the need for constant adjustment. They are generally designed to make it fairly easy for the beginner, though some require you to know the names of the brightest stars when setting them up.

The simplest are basically auto-tracking mounts, but the majority have Go To as well. Once set up, Go To mounts will find any object in their large database of planets, stars and deep-sky objects – though I strongly advocate getting to know the sky as well rather than relying on technology and indeed battery power. Most motorised altazimuth mounts can't easily be used without power.

Once you get into the larger sizes of telescope – say over 130 mm – you really need a motorised mount, though you can still buy 150 mm or even 200 mm instruments on basic equatorial mounts and add the motor later if you wish. Most mounts these days have Go To as a matter of course, however, and it's well worth having. If you plan to take long-exposure photos through your instrument an equatorial

Two Celestron 200 mm f/10 Go To Schmidt-Cassegrains. At left, the CPC 800 GPS, on altazimuth mount, for around £2250. This is good for visual observing and planetary imaging and is moderately portable. At right, the CGX Equatorial 800, with the same optical characteristics but on a heavier German-type equatorial mounting engineered for long-exposure astrophotography, and costing over £1000 more.

mount is usually necessary, though for planetary imaging an altazimuth instrument will be fine. But be warned that long-exposure astro-imaging requires a high level of complexity and involvement.

Some Go To mounts include cameras in their set-up system that identify the part of sky they are viewing and set the mount up for you automatically. And some now offer Wi-Fi control, allowing you to choose the object to be viewed using a smartphone app.

There is one type of altazimuth mount that is widely used for large, undriven reflectors, however – the Dobsonian. This was designed to give the maximum aperture at the lowest cost. To use it, you simply push and pull the instrument around the sky. These instruments are ideal for deep-sky observing at dark sites, but less so for studying planetary detail in towns, though if you know what you're doing you can get a large telescope for a comparatively small sum.

THE COST

Let's put some prices on the various types. Starting with refractors, you can get a 70 mm refractor on a basic altaz mounting for under £100. It's fine for viewing the craters of the Moon or the rings of Saturn, but the limited mounting makes more detailed observing tricky.

Between £150 and £300 (2018 prices) you can get a refractor up to 90 mm, a 100 mm catadioptric or a reflector up to 130 mm, maybe even including an auto-tracking mount. You'll need to find the objects for yourself, but such an instrument will show a wide range of objects of all types, given dark-sky conditions, as well as detail on the Moon and planets. You'll see more with larger

At the other end of the 200 mm scale is the Sky-Watcher Skyliner 200P f/6 Dobsonian – a Newtonian reflector on a basic but stable altazimuth mounting, for less than £350. This is best suited to visual observing of deep-sky objects, but it's also possible to get Dobs with optical encoders that will link to smartphones to help you find objects.

instruments, but this is all a matter of your priorities.

To get Go To and larger apertures you need to spend £500 to £750, for which sum you'll get a reflector up to 150 mm or a portable 127 mm catadioptric with a Go To mount. From there on upwards there are no limits! Typically, many people opt for a 200 mm Schmidt-Cassegrain on Go To altazimuth mount, which is a very usable and versatile instrument, for around £2500.

But if you are intent on getting a really good telescope with your retirement package, bear in mind a couple of things. One, you need to be quite dedicated to get the best out of it, and two, lugging a large telescope outside into the cold can require a lot of effort. So curb your enthusiasm for something that you can manage to start with. And remember the adage that most astronomers are very familiar with: the best telescope is the always the one you use the most, though it isn't always necessarily the largest!

Our view of the stars – a source of infinite amazement for scientists, stargazers and the millions of us who seek out rural places to rest and recuperate – is obscured by light pollution. It's a sad fact that many people may never see the Milky Way, our own Galaxy, because of the impact of artificial light.

LIGHT POLLUTION

Light pollution is a generic term referring to artificial light that shines where it is neither wanted nor needed. In broad terms, there are three types of light pollution:

• **Skyglow** – the pink or orange glow we see for miles around towns and cities, spreading deep into the countryside, caused by a scattering of artificial light by airborne dust and water droplets.

• **Glare** – the uncomfortable brightness of a light source.

• **Light intrusion** – light spilling beyond the boundary of the property on which a light is located, sometimes shining through windows and curtains.

The Campaign to Protect Rural England (CPRE) has long fought for the protection and improvement of dark skies, and against the spread of unnecessary artificial light. CPRE commissioned LUC to create new maps of Great Britain's light pollution and dark skies to give an accurate picture of how much light is spilling up into the night sky and show where urgent action is needed. CPRE also sought to find where the darkest skies are, so that they can be protected and improved.

MAPPING

The new maps are based on data gathered by the National Oceanographic and Atmospheric Administration (NOAA) in America, using the Suomi NPP weather satellite. One of the instruments on board the satellite is the Visible Infrared Imaging Radiometer Suite (VIIRS), which captures visible and infrared imagery to monitor and measure processes on Earth, including the amount of light spilling up into the night sky. This light is captured by a day/night band sensor.

The mapping used data gathered in September 2015, and is made up of a composite of nightly images taken that month as the satellite passes over the UK at 1.30am.

The data was split into nine categories to distinguish between different light levels. Colours were assigned to each category, ranging from darkest to brightest, as shown in the chart below. The maps

Colour bandings to show levels of brightness

Categories	Brightness values (in nw/cm²/sr)*
Colour band 1 (darkest)	<0.25
Colour band 2	0.25-0.5
Colour band 3	0.5-1
Colour band 4	1-2
Colour band 5	2-4
Colour band 6	4-8
Colour band 7	8-16
Colour band 8	16-32
Colour band 9 (brightest)	>32

The brightness values are measured in nanowatts/cm²/steradian (nw/cm²/sr). In simple terms, this calculates how the satellite instruments measure the light on the ground, taking account of the distance between the two.

are divided into pixels, 400 metres × 400 metres, to show the amount of light shining up into the night sky from that area. This is measured by the satellite in nanowatts, which is then used to create a measure of night-time brightness.

The nine colour bands were applied to a national map of Great Britain (see the following pages), which clearly identifies the main concentrations of night-time lights, creating light pollution that spills up into the sky.

The highest levels of light pollution are around towns and cities, with the highest densities around London, Leeds, Manchester, Liverpool, Birmingham and Newcastle. Heavily lit transport infrastructure, such as major roads, ports and airports, also show up clearly on the map. The national map also shows that there are many areas that have very little light pollution, where people can expect to see a truly dark night sky.

The results show that only 21.7 per cent of England has pristine night skies, completely free from light pollution (see the chart below). This compares with almost

57 per cent of Wales and 77 per cent of Scotland. When the two darkest categories are combined, 49 per cent of England can be considered dark, compared with almost 75 per cent in Wales and 87.5 per cent in Scotland. There are noticeably higher levels of light pollution in England in all the categories, compared with Wales and Scotland. The amount of the most severe light pollution is five times higher in England than in Scotland and six times higher than in Wales.

The different levels of light pollution are linked to the varying population densities of the three countries: where there are higher population densities, there are higher levels of light pollution. For example, the Welsh Valleys are clearly shown by the fingers of light pollution spreading north from Newport, Cardiff, Bridgend and Swansea. In Scotland, the main populated areas stretching from Edinburgh to Glasgow show almost unbroken levels of light pollution, creeping out from the cities and towns to blur any distinction between urban and rural areas.

Light levels in England, Wales and Scotland

Categories	England	Wales	Scotland	GB
Colour band 1 (darkest)	21.7%	56.9%	76.8%	46.2%
Colour band 2	27.3%	18.0%	10.7%	20.1%
Colour band 3	19.1%	9.3%	4.6%	12.6%
Colour band 4	11.0%	5.8%	2.8%	7.3%
Colour band 5	6.8%	3.8%	1.7%	4.6%
Colour band 6	5.0%	2.9%	1.2%	3.3%
Colour band 7	4.3%	2.1%	1.0%	2.8%
Colour band 8	3.2%	1.0%	0.9%	2.1%
Colour band 9 (brightest)	1.6%	0.2%	0.3%	1.0%

Adapted from Night Blight: Mapping England's light pollution and dark skies *CPRE (2016), with kind permission from the Campaign to Protect Rural England. To see the full report and dedicated website, go to http://nightblight.cpre.org.uk/*

MAP OF BRITAIN'S LIGHT POLLUTION AND DARK SKIES
COURTESY OF CPRE/LUC

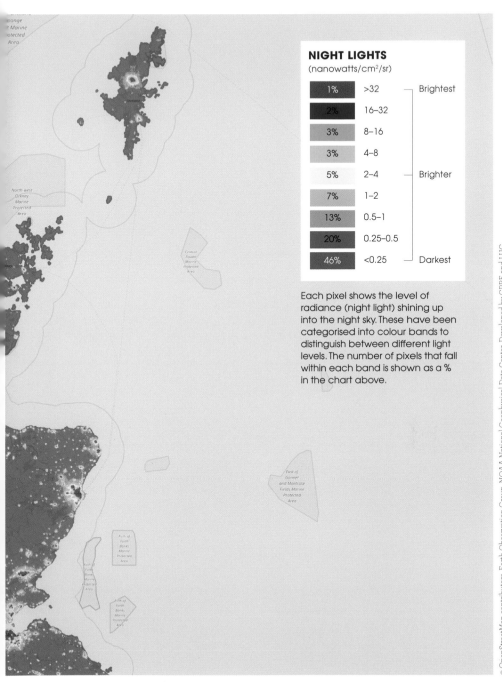

NIGHT LIGHTS

(nanowatts/cm²/sr)

1%	>32	Brightest
2%	16–32	
3%	8–16	
3%	4–8	
5%	2–4	Brighter
7%	1–2	
13%	0.5–1	
20%	0.25–0.5	
46%	<0.25	Darkest

Each pixel shows the level of radiance (night light) shining up into the night sky. These have been categorised into colour bands to distinguish between different light levels. The number of pixels that fall within each band is shown as a % in the chart above.

© OpenStreetMap contributors, Earth Observation Group, NOAA National Geophysical Data Center. Developed by CPRE and LUC.

DARK SKIES & LIGHT POLLUTION

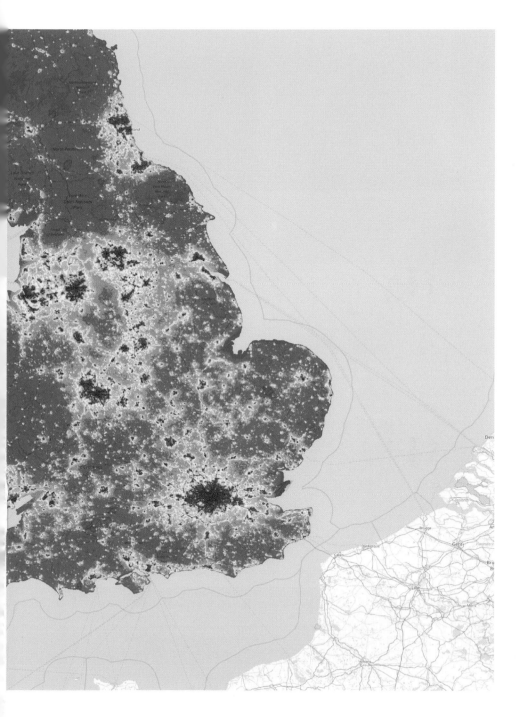

Heather Couper and **Nigel Henbest** – both Fellows of the Royal Astronomical Society – are internationally recognised writers and broadcasters on astronomy and space. As well as writing more than 40 books and 1000 articles, they founded an independent TV production company specialising in factual programming.

Heather is a past President of both the British Astronomical Association and the Society for Popular Astronomy. She is a Fellow of the Institute of Physics and a former Millennium Commissioner, for which she was awarded the CBE in 2007.

After researching at Cambridge, Nigel became a consultant to both *New Scientist* magazine and the Royal Greenwich Observatory, and edited the *Journal of the British Astronomical Association*. He is a future astronaut with Virgin Galactic.

ACKNOWLEDGEMENTS

PHOTOGRAPHS
Front cover: Sara Wager; The Whirlpool Galaxy.
Celestron: 87 (bottom), 88.
Galaxy Picture Library: Ed Cloutman 53; Alan Dowdell 47; Peter Jenkins 29; Steve Knight 17; Damian Peach 41; Dave Ratledge 71; Robin Scagell 23, 87 (top); Nik Szymanek 59, 77; Pete Williamson 1, 11, 35, 65.
NASA: /JPL-Caltech/UCLA 2–3; ESA, J. Clarke (Boston Univ.) and G. Bacon (STScI) 18; JPL 30; Carnegie Institution of Washington 31; JPL-Caltech 37; Jerry Lodrigruss (Catching the Light) 54; JPL/University of Arizona 55; JPL/Caltech 67; VegaStar Carpentier 83; Yuugi Kitahara 84 (top); Robert Gendler 84 (bottom); JPL-Caltech/STScI 85.
Optical Vison Ltd: 86, 89.

Wikimedia: The High Fin Sperm Whale 7; Antonio Fernandez-Sanchez 24; David Moug 36.

ARTWORKS
Star maps: Will Tirion/Philip's with extra annotation by Philip's
Planet event charts: Nigel Henbest
Pages 90–95: Adapted from *Night Blight: Mapping England's light pollution and dark skies* CPRE (2016), with kind permission from the Campaign to Protect Rural England.
To see the full report and dedicated website, go to http://nightblight.cpre.org.uk/
Maps © OpenStreetMap contributors, Earth Observation Group, NOAA National Geophysical Data Center. Developed by CPRE and LUC.